UNMANNED COMBAT AERIAL VEHICLES:

EXAMINING THE POLITICAL,

MORAL, AND SOCIAL IMPLICATIONS

SCHOOL OF ADVANCED AIR AND SPACE STUDIES

DISCLAIMER

The conclusions and opinions expressed in this document are those of the author. They do not reflect the official position of the US Government, Department of Defense, the United States Air Force, or Air University.

ABOUT THE AUTHOR

Lt Col James "Opie" Dawkins, Jr, received his commission through Officers Training School in 1989. He has logged over 2200 hours in the F-111, F-16, and the B-2. Lt Col Dawkins' past operational assignment include: Weapons Instructor, F-111s, Cannon AFB, NM; Flight Commander, F-16CJs, Spangdahlem AB, Germany; Chief of Wing Weapons, B-2s, Whiteman AFB, MO; and Operations Officer, Global Strike Branch, J-3, The Joint Staff, Pentagon, Washington D.C.

Lt Col Dawkins is a 2004 graduate of the US Army Command and General Staff College and a 1994 distinguished graduate of the US Air Force Weapons School. His education includes a bachelor's degree in Business from Baylor University, in 1988 and a Masters of Aeronautical Studies from Embry Riddle University, in 2000. In July 2005, he will begin B-2 requalification training at Whiteman AFB. Lt Col Dawkins is married to the former Daina Marie Duchaine of Coco, FL. They have a daughter, Jacqueline, 16, and a son, Jimmy, 12.

ACKNOWLEDGMENTS

I wish to thank Col(S) Scott Gorman and Dr. Gary Schaub for their help in guiding me through this journey. They performed a great service as they patiently listened and read through my many drafts, never failing to offer timely advice and greatly needed course corrections. I also wish to thank the faculty of the School of Advanced Air and Space Studies as they provided a needed sounding board early in my quest for a worthy thesis topic.

Most importantly, I wish to thank my family for supporting my academic aspirations. As of July 2005, they will have endured four PCS moves within three years. No words can express my sincere appreciation for their love, patience, and understanding during those moves and during the times I was absent in spirit, off struggling with this paper in the corner of my bedroom office.

ABSTRACT

There will likely be political, moral, and social implications of UCAV employment that strategists and military commanders will need to pay attention to as they craft strategies for future conflict. UCAVs are a very appealing option for the politician faced with use-of-force decisions due to reduced forward basing requirements and the possibility of zero friendly operator casualties. The flexibility of the weapon system offers the politician a seemingly high degree of control over the process of war. Together, these advantages may make a politician more inclined to use force first rather than last. In the moral realm, UCAVs are neither immoral nor illegal simply because risk to one of the combatants is removed. Additionally, notions of chivalry and fairness are not good standards by which to judge this technology. The social impact of widespread UCAV employment on the operator is an area of further concern. Remote-control war, however, does not change the underlying assumptions that have been the basis for the military ethos in the past. The final chapter highlights the dynamic between political, moral, and social issues as it addresses a range of possible unintended consequences resulting from extensive UCAV employment. Ultimately, the purpose of this thesis is to provide strategists greater clarity on the political, moral, and social issues surrounding UCAV employment. Doing so allows them to more effectively address, both objectively and subjectively, the implications of this new technology.

CONTENTS

Introduction

We're going to tell General Atomics to build every Predator they can possibly build.

Air Force Chief of Staff, General John Jumper, testifying before the Senate Armed Services Committee in 2005

Unmanned aerial vehicles (UAVs) have greatly contributed to America's success in the current war on terrorism. While designed primarily as an intelligence, surveillance, and reconnaissance platform, the Predator MQ-1 has evolved into an effective weapon system, able to carry out pinpoint strikes in dynamic battlefield conditions. One of the most well-known and highly publicized strikes occurred on 3 November 2002. On that day in the desert of Yemen, a black sports utility vehicle carrying six al Quaeda operatives suddenly exploded. Ground observers near the scene believed it was merely another car bomb, a not so uncommon occurrence in the region. Soon after, however, assistant Secretary of Defense, Paul Wolfowitz, boasted that this was the work of a CIA operated Predator UAV using a Hellfire precision missile.[1]

The Air Force has also used weaponized Predators extensively in Afghanistan and Iraq, logging more than 80,000 flying hours since the beginning of operations in 2001.[2] Targets have ranged from suspected terror hideouts in the mountains and urban areas to satellite communication infrastructure in downtown Baghdad. Predator operators have even attempted to shoot down airborne assets. For instance, prior to Operation Iraqi Freedom (OIF), the Air Force used a Predator armed with a Stinger air-to-air missile to engage an Iraqi Mig-25 that flew into the southern no-fly zone. Though unsuccessful – the Mig shot down and destroyed the Predator – it does show how serious the Air Force is about this new capability.[3]

The Hellfire-equipped MQ-1 Predator, however, is just the start of an extensive DoD-wide unmanned weapons procurement plan. DoD has invested over $4 billion in UAV development, procurement, and operations. By 2010, the department is expected to

---

[1] Evan Thomas and Mark Hosenball, "The Opening Shot," *Newsweek,* 18 Nov 2002, 48.
[2] Bryan Bender, "Attacking Iraq from a Nevada Computer," *The Boston Globe,* Sunday, 3 April 2005, sec. A6.
[3] Lt Col Gary Fabricus, Squadron Commander of one of the first two Weaponized Predator Squadrons, interviewed by author 14 January 2005.

invest at least $10 billion more.[4] On the heels of successful offensive operations in Afghanistan and Iraq using weaponized Predator UAVs, the DoD is actively pursuing new unmanned systems built from the ground up for offensive strike operations. These weaponized UAVs are now referred to as unmanned combat aerial vehicles, or UCAVs.

According to DoD Joint Publication 1-02, a UCAV is "a powered, aerial vehicle that does not carry a human operator, uses aerodynamic forces to provide vehicle lift, can fly autonomously or be piloted remotely, can be expendable or recoverable, and can carry a lethal or non-lethal payload."[5] Although UCAVs and cruise missiles are both considered unmanned aircraft, UCAVs differ from cruise missiles in two distinct ways. First, operators can recover UCAVs at the end of their flight, quickly refueling and rearming them for subsequent operations, something that can not be done with the cruise missile. Second, unlike cruise missiles that generally have a single warhead tailored and integrated into their airframe, current and future UCAVs have the capability to carry multiple, independently targetable munitions.[6]

UCAVs are extremely flexible due to their persistence, wide range of sensors, diverse weapons load, and command and control capabilities. The MQ-1 and follow on MQ-9 Predators have an endurance capability in excess of 24 hours.[7] Future UCAVs will have even greater duration capabilities. No longer will operations be limited by crew fatigue concerns. Instead of relying on a single crew to fly long duration attack missions, UCAVs operations allow multiple crews to take turns during the course of a single mission. Fresh operators can be rotated into the control center at mission critical times such as during target engagements thereby reducing the chances of mistakes, subsequently increasing mission effectiveness. In today's environment, having the ability

---

[4] US Department of Defense, *Unmanned Aerial Vehicles Roadmap: 2002-2027,* (Washington, D.C.: Office of the Secretary of Defense, December 2002), iv. "The document presents the Department of Defense's Roadmap for developing and employing UAVs and UCAVs over the next 25 years. The goal of the document is to define clear direction to the Services and Departments, in concert with the Defense Planning Guidance for a logical, systematic migration of mission capabilities to a new class of military tools. It describes the Service's ongoing UAV efforts and includes a description of the future potential of UAVs. The roadmap is to be the directive in cross-program areas such as standards development and other interoperability solutions."

[5] US Department of Defense, *Joint Publication 1-02: Dictionary of Military and Associated Terms*, 30 Nov 2004.

[6] DoD UAV Roadmap, 2.

[7] DoD UAV Roadmap, 8-10.

to loiter over designated areas for many hours engaging pre-planned and pop-up targets is extremely beneficial.

Additionally, the UCAV's impressive sensor capabilities allow for built-in reconnaissance and battle damage assessment. Current UCAVs, like the Predator, use various combinations of high definition electro-optical and infrared sensors that enable the aircraft to turn darkness into daylight. Future systems will have this same capability and also incorporate multispectral, hyperspectral, and synthetic aperture radar sensors that will enable operators to see through inclimate weather and dense forest or canopy. In addition, future magnification improvements will allow operators to identify individuals through facial recognition technology.[8] These capabilities are vital in conflicts where adversaries use tactics that focus on concealment and constant movement.

Robust communications capabilities allow operators and observers the ability to monitor events in real time. UCAVs can be flown from almost any location on the earth through the use of satellite communications. Furthermore, the imagery obtained from their sensors can be directed to the operators on the ground or in the air, the military commander in the command center, or even the White House if desired.[9]

Admittedly, current UCAVs like the Predator are not impervious to ground and airborne threats such as surface-to-air missiles, anti-aircraft-artillery fire, and enemy fighters. The current version of the Predator is not stealthy, flies slow and at low altitudes, and has no on-board self protection measures such as chaff, flares, or electronic jamming pods. However, the Air Force is incorporating the latest stealth technology as well as other self-protection measures similar to the ones used in current aircraft into the X-45.

The MQ-9, the Predator's bigger brother, is the first unmanned aircraft designed and built specifically for weapons delivery. While the first weaponized Predator only carried two Hellfires, the MQ-9 Predator can carry up to ten of these weapons using its 3500 pound payload capability. The stealthy X-45 will have an even larger 4500 pound payload that will enable it to carry a wider range of weapons, like the 2000 pound JDAM

---

[8] DoD UAV Roadmap, 55.
[9] DoD UAV Roadmap, 39-40.

penetrator.[10]   The Air Force has budgeted for 36 advanced, stealthy X-45 UCAVs to be delivered by 2010, a few of which are currently flying test missions at Edwards AFB.[11]

Even before the events of September 11, 2001, these systems received high-level support. The 2001 National Defense Authorization Act included a provision which directed that "the DoD aggressively develop and field unmanned combat systems in the air and on the ground so that within 10 years, one-third of our operational deep strike aircraft would be unmanned."[12]   Regardless of forecast numbers, UCAVs are destined to be a major asset in the future Air Force arsenal.

The Air Force is now preparing for large-scale unmanned aircraft operations. Currently, the Air Force conducts stateside Predator operations from Indian Springs Airfield in Las Vegas, Nevada. In the past, this small airfield was designated as an auxiliary airfield to be used only for emergencies by manned aircraft and Predator launch and recovery operations. Consequently, it had little in the way of base support facilities. In light of the growing unmanned fleet, however, the Air Force is transforming the field into a fully functioning Air Force base and has established the first dedicated Predator Air Wing designating it as a "center of excellence" for unmanned vehicle operations.[13] Additionally, the Air Force intends to open up 12 new Predator squadrons of 12 aircraft each. By 2007, Texas and Arizona should have their own squadrons, followed in 2009 by New York.[14]   The age of unmanned warfare is upon us.

Fielding and employing the next generation of UCAVs requires DoD to do more than just follow the standard procurement process of R&D, purchase, tactics development, and fielding. Placing the operator in a ground control trailer thousands of miles from the battlefield, in a benign and risk free environment, calls for the military to take a deeper look into the effects of such an action. Noted authors have already addressed a few of the possible implications stemming from UCAV employment and the notion of risk-free warfare. Taken at face value, these implications may give Western

---

[10] DoD UAV Roadmap, 10-12.

[11] DoD UAV Roadmap, iv.

[12] Senate Armed Services Committee, *National Defense Authorization Act for Fiscal Year 2001,* 106th Cong., 2nd session, 2001, section 217.

[13] Maj General Stephen Goldfein, "Air Force stands up UAV Center of Excellence," press conference, Nellis Air Force Base, Nev., 17 Mar 2005, available on-line at airforcelink.com.

[14] The Air Force has yet to publicly release the specific locations of these new units. Peter Pae, "Air Force Wants Big Boost in Predator Fleet," *Los Angeles Times*, 19 March 2005.

society great cause for concern. Michael Ignatieff, in his book, *Virtual War*, argues that waging risk-free war changes war into something "that ceases to be fully real." This, in turn, may tempt politicians (with the support of their constituents who also see modern conflict as a video game) to resort to war more often. Ultimately, Ignatieff concludes, the public and the military must not think they can conduct war without risk – they must be willing to get their "hands dirty" before they can "do what is right."[15] Army Chaplain Keith Shurtleff, in his article *The Effects of Technology on Our Humanity*, agrees, adding that clean wars are also more sustainable because a powerful deterrent to warfare is now removed. He asserts that "the benefit of keeping war in the realm of the horrifying can be found in the very deterrence such horror provides. Conversely, disengagement from those horrors may have the consequence of making war more acceptable, and therefore more prevalent."[16]

Others are concerned about the effect that UCAV employment will have on the operators of these systems. Air Force legal expert Charles Dunlap argues in his monograph, *Technology and the 21st Century Battlfield: Recomplicating Moral Life for the Statesman and the Soldier*, that this type of warfare challenges the foundational basis of the military ethos. He asserts that "statesmen and soldiers should not assume that such combatants will automatically share the military's traditional values that restrain illegal and immoral conduct in war ... a new ethic will need to emerge in order to guide the behavior of this new breed of 'console warriors'."[17]

But should this new method of warfare really concern us, or is this merely the unwarranted concern of just a few? Answering this question is the purpose of this work. My thesis is simple: There are political, moral, and social implications of UCAV employment to which strategists and military commanders need to pay attention as they craft strategies for future conflict.

The paper begins with an examination of the political implications of UCAV employment. UCAVs are a very appealing option for the politician faced with use-of-

---

[15] Michael Ignatieff, *Virtual War: Kosovo and Beyond*, (New York, NY: Metropolitan Books, 2000), 161, 215.

[16] D. Keith Shurtleff, "The Effects of Technology on Our Humanity," *Parameters,* Summer 2002, 103.

[17] Charles J. Dunlap Jr., *Technology and the 21st Century Battlfield: Recomplicating Moral Life for the Statesman and the Soldier*, (Carlisle PA: Strategic Studies Institute, U.S. Army War College, 1999), 30.

force decisions due to reduced forward basing requirements and the possibility of zero friendly operator casualties. Moreover, the flexibility of the weapon system offers the politician a seemingly high degree of control. Together, these advantages may make a politician more inclined to use force first rather than last. The extent to which this occurs, however, depends on the type and nature of the conflict.

The paper next turns to the issue of morality and argues that UCAVs are neither immoral nor illegal simply because risk to one of the combatants is removed. The second chapter uses currently accepted principles, rules, and norms that govern the weapons of war to analyze the morality of the UCAV. Additionally, the chapter argues that notions of chivalry and fairness are not good standards by which to judge this technology.

The social impact of widespread UCAV employment on the operator is another area of concern. UCAVs in general are an awkward fit with what society and the military associates with warrior culture or ethos. Remote-control war, however, does not change the underlying assumptions that have been the basis for the military ethos in the past. Though a new military ethic is not needed for these console warriors, the way military leaders stress the ethos needs to change as technology continues to increase the distance between combatants.

The first three chapters address the political, moral, and social implications of UCAV employment separately and in isolation. However, the value of this thesis lies in understanding the interrelationships between the three. In warfare, political, moral, and social issues are complex and interdependent, with each one impacting the other in an interactive process. The final chapter highlights this dynamic as it addresses a range of possible unintended consequences resulting from extensive UCAV employment. UCAVs, for instance, may increase the occurrence of war or even increase the overall danger to non-combatants in a process the thesis calls risk transfer, an occurrence that has taken place countless times between combatants as they have sought to asymmetrically counter their enemy.

The purpose of this thesis is to provide strategists greater clarity on the political, moral, and social issues surrounding UCAV employment. Doing so allows them to more effectively address, both objectively and subjectively, the implications of this new technology. According to General Jumper, future UCAV employment requires us to "think it through and get it right."[18] Getting it right requires understanding the implications of UCAVs in all three areas – political, moral, and social – lest the military field a system that is ultimately politically, morally, and/or socially unusable.

[18] Tirpak, John A. "Airpower and 'The Long War', Four Star Forum: Eyes on the Future," *Air Force Magazine,* November 2004, 87, 79.

Chapter 1
Political Implications of UCAVs

UCAVs offer U.S. political leaders many advantages. Removing the human from the cockpit reduces both friendly casualties and forward basing requirements. Moreover, the UCAV's inherent persistence, adaptive sensor suite, and wide range of weapons make it an extremely flexible weapon. But will these advantages make politicians more inclined to resort to force first rather than last?

Reduced Forward Basing

Future UCAVs will reduce the amount of forward basing needed to support operations. Admittedly, the Air Force's current UCAV, the Hellfire-armed Predator, does not remove the requirement for forward basing of some support personnel and even a few pilots. Although operators can control them from bases in the United States, a small support contingent of maintenance personnel and aircrew is still required to launch and recover the aircraft at forward operating locations. Additionally, its slow speed and lack of refueling capability requires that it be stationed just as close to the battlefield as some tactical aircraft. Later generations of UCAVs, however, such as the X-45, will have increased speed, payload, and endurance which will allow for these personnel to be further removed from the area of operations. Furthermore, future in-flight refueling capability will further diminish the need for forward support personnel, allowing the military to launch and recover UCAVs from the continental United States.[19] Absent in-flight refueling capability, the military has already proven the capability to conduct long-range unmanned aerial operations. In 2001, the Global Hawk reconnaissance UAV set a precedent when it flew 7,500 miles non-stop from the United States to Australia.[20]

Reduced forward basing requirements translate into less interference from other nations during international conflicts. In many scenarios, the United States had to rely on allies or friendly nations to provide airfields and staging areas for combat forces due to the short range of many tactical aircraft. At times this can cause problems when allies

---

[19] DoD UAV Roadmap, 53.
[20] Air Force Link Website, Global Hawk, www.af.mil/factsheets/factsheet.asp?fsID=175

place limitations on U.S. operations, reducing the military's capability to effectively employ airpower. The extensive coordination necessary for operations requiring the use of foreign bases tends to complicate the situation for the political leader. UCAVs will allow the politician a great deal of flexibility as coordination with foreign nations can be reduced or eliminated altogether.

The nature of the conflict, however, will determine how important this advantage is in the politician's decision calculus. Smaller footprint and reduced forward basing requirements will be very appealing to politicians faced with conflicts lacking international support. Generally, low international support translates into reduced forward basing options.[21] For example, both the Turkish and Saudi Arabian governments have in the past imposed limitations on the numbers of aircraft and kinds of missions they allow to be based and flown from their airfields. During Operation Southern Watch (OSW) and later, Operation Iraqi Freedom (OIF), Saudi Arabia did not allow the United States to use its airfields to conduct offensive strikes, allowing only tankers and other support aircraft to use its strategically-located bases.[22] Likewise, Turkey, during OIF, denied the United States the use of its ground and air bases. The lack of a northern front gave the Iraqis a small sanctuary during the initial stages of the war, making post-war security and stability more complicated.[23]

Even close allies impose limits on what can and cannot be accomplished from their airfields during unpopular wars. Before executing the 1986 attack against Libya, President Reagan had to secure permission from the British government in order to use F-111s stationed at English airbases. He also asked France for over flight permission, something they subsequently denied. Their denial added 1,300 miles and seven hours to an already long flight, greatly increasing the amount of aerial refueling support required for the mission. Moreover, the tactical success of the mission suffered as the long flight pushed the pilots and their aircraft past the bounds of their normal operating conditions.[24] More recently during OIF, the British government restricted U.S. bombers based in

---

[21] David A. Shlapak, et al, *A Global Access Strategy for the U.S. Air Force,* (Santa Monica, CA: RAND Corporation, 2002), 15-43.
[22] Barbara Starr, U.S. to move operations from Saudi base, CNN.com, April 29, 2003, online at www.cnn.com/2003/world/meast/04/29/sprj.irq.saudi.us/
[23] Max Boot, "The New American Way of War," *Foreign Affairs*, July/August 2003, online at www.foreignaffairs.org/20030701faessay15404-p10/max-boot/
[24] Shlapak, 8-9.

Britain and at their remote Indian Ocean island of Diego Garcia from attacking certain targets without first receiving approval from the British government.[25] Long-range, aerial refuelable UCAVs that operate from U.S. bases will alleviate many of these problems.

Conversely, in conflicts that are easily justified, explained, and supported by the international community, the UCAV's smaller footprint and reduced forward basing requirements may not offer the politician a decisive advantage over current platforms. In cases like these, allies have generally been willing to open up their bases for American combat operations. In World War II, obtaining forward operating bases for B-25s and B-17s and various fighter aircraft did not prove very difficult because the United States was helping its allies fight a war of national survival. Likewise, in the days immediately after September 11th, many nations opened up their bases to bombers, fighters, and airlift aircraft to accommodate the United States in its air and ground operations in Afghanistan.[26]

Therefore, in conflicts that international audiences support and understand, the UCAV's smaller footprint may not weigh heavily in a politician's use of force decision. In these cases, politicians can generally count on allies or other friendly nations to offer the United States the use of their bases. In conflicts that the international community is hesitant to support, however, the UCAV's reduced forward basing requirements are a great advantage that may make politicians more prone to resort to force first, or at the very least, without seeking international consensus on an issue.

Reduced Casualties

The most obvious advantage of the UCAV derives from its inherent ability to reduce friendly casualties, both directly and indirectly. Humans, once required to pilot planes from the cockpit, can now do so from the safety of a ground control center, thousands of miles from the battlefield. Arguably, modern cruise missiles, launched from the sea or from the air, have a similar capability. However, the very nature of missions

---

[25] Maj Gil Petrina, B-2 Mission Planner during OIF, interviewed by author, 4 Mar 2005.
[26] Congressional Research Service, *Operation Enduring Freedom: Foreign Pledges of Military & Intelligence Support,* Report for Congress, October 17, 2001, online at http://fpc.state.gov/documents/organization/6207.

that employ these weapons usually involves a certain degree of risk for the operators. For instance, submarines and surface ships equipped with cruise missiles, though often operating in the relative safety of international waters, are subject to enemy attack as well as mechanical malfunctions and operational accidents that can kill or injure those on board. The recent accident involving the USS San Francisco provides one example of this. During training operations in the Pacific, the submarine hit an uncharted underwater mountain resulting in one fatality and numerous injuries.[27] Likewise, manned aircraft missions also have a certain degree of mission risk or opportunity cost inherent in each flight. This cost or risk is always present regardless of how far the combatants are from the enemy. UCAV operations, however, reduce these costs to near zero.

The idea of zero operator deaths is very appealing to the politician responsible for making decisions requiring the use of force. With domestic audiences, friendly casualties are one of many important factors with which the politician contends and support in this area can be difficult for the politician to obtain. The degree of difficulty in obtaining this support, however, often depends on the nature of the conflict.

One-time strikes or raids are where the politician may find the UCAV most appealing. Short term operations are very appealing to the politician regardless of what type of assets the military uses. The short term nature of these strikes and the small number of forces required means reduced chances of large numbers of casualties. Moreover, post-strike media attention can only focus on past events, not future problems associated with conflicts that are longer in duration. Furthermore, there is less of a requirement for politicians to garner support from Congress or the public for continuing an operation that is already over.[28]

Prisoners of war resulting from downed manned aircraft, however, can turn a short and straightforward strike, both in terms of military and political considerations, into an enduring event. [29] POWs are a liability in any type of conflict and can greatly complicate matters for the politician. POW negotiations extended the Korean War for at least a year. Additionally, though not technically prisoners of war, the crew of the

---

[27] William H. McMichael, "Punishment Meted Out for Six in Grounding of Submarine," *Navy Times.com,* 22 March 2005.
[28] Woodley, 32-49.
[29] Woodley, 49.

damaged EP-3 that landed in China in April of 2001 provided the Chinese government a great deal of leverage. Likewise, had Capt Scott O'Grady been captured by enemy forces after he was shot down in Bosnia in June of 1995, this would have greatly complicated an already divisive political situation for President Clinton in the Balkans. UCAV employment not only reduces concerns over friendly casualties, but it eliminates prisoner of war (POW) concerns as well. In some cases this may be more important to the politician than casualties.

UCAVs are also attractive to the politician in politically sensitive conflicts. The public has been historically unwilling to accept the costs of casualties in these "complex political situations characterized by civil conflict, in which U.S. interests and principles are typically much less clear and success is often elusive at best."[30] Somalia, Haiti, Bosnia, and Kosovo are examples of operations that fall into this category.

In Kosovo, Congress and the Executive Branch often fought over whether or how deeply the United States would involve itself. The White House had to engage in a robust public relations campaign in order to sell the intervention to the American people, a campaign that started with President Bush's "Christmas warning" to Milosevic on Christmas Eve in 1992.[31] But even after the Executive Branch reached a degree of consensus, friendly casualty concerns greatly impacted future operations. Concerned that even low levels of friendly casualties would undercut public support, NATO leaders implemented stringent rules of engagement for air operations, limiting aircrews from descending below 15,000 feet for fear that doing so would increase the chance that an aircraft would be shot down. They also removed the possibility of casualties resulting from ground combat, publicly stating early on in the war that ground intervention options were off the table.[32] Using an air campaign to coerce Milosevic was a very attractive option for the political leaders of the United States and NATO because of the lower probability of friendly casualties. The use of UCAVs alone would have been an even more appealing option.

---

[30] Larson, xvii.
[31] R. Ross Woodley, "Unmanned Aerial Warfare, Strategic Help or Hindrance," (master's thesis, School of Advanced Air and Space Studies, Air University, Maxwell Air Force Base, AL, 2000), 54.
[32] Joe Lockhart, White House Press Briefing on Operation Allied Force, May 21, 1999, online at www.clintonfoundation.org/legacy/052199

Conversely, in conflicts that are easily justified to a domestic constituency, like World War II, Korea, and Afghanistan, reduced casualties from UCAV employment may not offer the politician much of an advantage over manned aircraft. Those who believe otherwise likely do so out of concern that the public is extremely casualty averse in all conflict situations. However, according to Eric Larson's 1996 Rand Report, *Casualties and Consensus: The Historical Role of Casualties in Domestic Support for U.S. Military Operations*, such generalizations are without merit. His study addressed the question of just how tolerant the American public is to casualties in war. He argues that:

> Majorities of the public have historically considered the potential and actual casualties in U.S. wars and military operations to be an important factor in their support, and there is nothing new in this. But the current attention to the public's unwillingness to tolerate casualties misses the larger context in which the issue has become salient: The simplest explanation consistent with the data is that support for U.S. military operations and the willingness to tolerate casualties are based upon a sensible weighing of benefits and costs that is influenced heavily by consensus (or its absence) among political leaders. In short, when we take into account the importance of the perceived benefits and several other factors, the evidence of a recent decline in the willingness of the public to tolerate casualties appears rather thin.[33]

World War II was a lengthy war, yet once the public realized how important the stakes were, they were more than willing to accept a very high level of casualties, quickly and persistently supporting the political leaders throughout the entire course of the war, even in the aftermath of battles with high casualties. The early days of Korea also showed a public willing to accept high levels of casualties for a cause that was easily explained by politicians and understood by the public.[34] Casualty aversion, therefore, is not an overriding factor in all decisions to use force.

This willingness to sustain casualties has not changed in modern times. Immediately after September 11th, President Bush asked the nation to support combat operations to remove Taliban control over Afghanistan. A Gallup poll conducted two months later showed that 80% of the public supported the use of ground troops in

---

[33] Eric V. Larson, *Casualties and Consensus: The Historical Role of Casualties in Domestic Support for U.S. Military Operation,* (Santa Monica, CA.: RAND Corporation, 1996), xv-xvi.
[34] Larson, xvi-xvii.

Afghanistan while acknowledging their involvement would mean casualties.[35] Furthermore, even though over 1000 military service members had been killed serving in Iraq at the time of the 2004 U.S. Presidential elections, a majority of the American public voted to reelect the man responsible for taking the country to war.

In sum, in conflicts where national interests are at stake, UCAVs do not offer the politician a significant political advantage over manned aircraft. This all changes, however, during politically sensitive conflicts, where friendly casualties and POW concerns are an overriding consideration. Armed with these advantages, the president may be tempted to forgo debate on the use of force, bypassing other instruments of power.

Increased Control

UCAVs give politicians a feeling of control that they do not otherwise experience with other aerial assets. This control comes in many forms. First, the politician can conduct operations without having to secure forward basing permissions, thereby controlling the timing and tempo of operations absent interference from host nations. Second, the politician does not have to worry about incurring casualties during an operation, in effect allowing him a large degree of control over the debate in Congress and with the public over decisions requiring the use of force. If he knows that an action is risk-free, then he is unlikely to feel compelled to vet an issue through Congress or use other means of dealing with the situation. Finally, the politician may believe that he can precisely control operations all the way down to the tactical level. From takeoff to landing, they may be tempted to get involved in the details of target and weapons selection, go/no-go criteria, and weapons release authorization in a way only depicted in Tom Clancy novels. Having control over these three areas may lead the politician to believe that they can precisely control the outcome of an operation, more closely tying together diplomacy and force, allowing them to ratchet up or down coercive pressure as needed.

---

[35] David W. Moore, "Eight of 10 Americans Support Ground War in Afghanistan," *The Gallup Organization,* 1 November 2001, on-line at www.gallup.com/content/login.aspx?ci=5029.

This feeling of control, however, may prove to be dangerously seductive. Though presidents have every right to involve themselves in the details of military operations, it may not always be desirable. President Ford discovered this during the Mayaguez incident. On May 12, 1975 Khmer Rouge forces seized the *SS Mayaguez* cargo ship on its way from Hong Kong to Thailand, subsequently capturing its crew. Word of the incident quickly made its way to the White House where President Ford, facing his first military action as president, directed that the military make every attempt to rescue the crew before the enemy forces reached land. The President knew that a sea rescue was fraught with less peril than one undertaken on land due to the difficulty of finding the men in the dense jungle.[36]

Under direction from the President Ford, forces from the U.S. Support Activities Group/7th Air Force (USSAG/7AF) at Nakhon Phanom Royal Thai Air Base began flying armed reconnaissance missions over the area. F-111s, A-7s, F-4s, U-2s, and EC-130s (airborne command post) kept constant watch over the unfolding crisis. A robust command and control radio-relay infrastructure allowed strategic decision makers to get real time updates on the situation from operational and tactical level forces. Through a direct and continuous communications link between aircraft overhead the area of interest and the White House Situation Room, the President had the ability to talk directly to the pilots or ground commanders during the operation if he desired, which, as is widely reported, President Ford did during the mission. Though not technically true – Ford did not talk *directly* to the pilots –the President did receive up to the minute reports from Brent Scowcroft, his National Security Advisor as Scowcroft monitored in-flight communications and talked with commanders on the scene from the White House Situation Room.[37]

President Ford, whether he wanted to or not, was intimately involved in the operational and tactical details of the operation. During one tense moment, the President was called on to make a decision on whether or not an A-7 aircraft would be allowed to disable the rudder on the enemy ship carrying the captured crew by strafing the aft section of the ship. Commanders on the scene believed that this action afforded the best

[36] Ralph Wetterhahn, *The Last Battle: The Mayaguez Incident and the End of the Vietnam War,* (New York, NY: Carroll & Graf Publishers, Inc., 2001), 25-35.
[37] Wetterhahn, 95-96.

chance to rescue the crew before the ship reached land. While pilots orbited awaiting his answer, the President launched into a heated discussion with his senior military and political advisors over the location and speed of U.S. Navy ships steaming to the area. When General Jones, the acting Joint Chiefs of Staff, told him that the aircraft carrier Coral Sea was at flank speed, "making twenty-five knots," Ford quickly disagreed and stated that, "Flank speed is thirty-three knots." Jones replied that the Navy commander on the scene had told him that that was the best the ship could do under the conditions in the area. Then the discussion turned to how close the aircraft carrier needed to be in order to begin conducting air operations. Senior leaders, including the president, found themselves discussing the number of available catapults, best wind direction and speed for launch and recovery operations, launch and recovery cycle times, and other details. Meanwhile, the pilots waited for an answer, with the ship getting closer and closer to the shore as each minute passed.[38]

President Ford found himself caught up in the drama, with his emotions prevailing over sound judgment. In the end, based on concern that some of the captured crew might die during the strafing, he decided to drop tear gas on the ship; however, this did not stop it from reaching the shore.[39] The President was then left with no other option than to mount an air and ground assault on the island where the men were now being held. In the end the Khmer Rouge relented and ceded to U.S. demands to release the crew. During the operation 41 marines died in order to save 40 crew members. Three were left behind on the island, their bodies left un-recovered until the mid-1990s.[40] Arguably, had President Ford left the decision to disable the ship with a strafing mission to the commander on the scene, the ground invasion would not have been necessary.

Today, the president still has the ability to directly communicate his desires to the pilots of manned as well as unmanned aircraft. There is a distinct difference, however, between the two types of aircraft in this regard. With UCAVs, the president does not have to worry about placing the operator in harms way. Removing this concern seemingly offers the president even greater control over unfolding events. Having control over this one aspect, however, is not enough. War, by its very nature, is

---

[38] Wetterhahn, 97-99.
[39] Wetterhahn, 103-109.
[40] Wetterhahn, 257-263.

inherently complex. Uncertainty abounds despite our desire for clarity and control. And regardless of man's attempts to use technology to simplify and clear the fog that surrounds it, war will remain uncontrollable. Yet the temptation will always be there.

One downside of this increased ability to control the means of war is that it tempts strategic level decision makers to engage themselves in the details of an on-going operation. If they do so, it may become difficult for them to remain focused on the strategic objectives. They can easily lose sight of the larger ramifications of how actions impact the international community. An operation using UCAVs would likely be just a small part of a larger strategy that includes diplomatic, economic, and informational aspects. The hypnotic feeling of control derived from UCAV employment may tempt them to forgo other instruments of power, instead, relying solely on the use of force for all disputes even when force may not be the best course of action. Every dispute might begin to look like a nail for the president's new-found hammer.

Moreover, by directly engaging in tactical aspects of the operation, the president becomes directly linked to mission successes and more importantly, mission failures. Plausible deniability is lost. Blame for failure rests squarely on his shoulders and it is more difficult for him to shift blame to those men who serve him. The hypnotic feeling of control that UCAVs seemingly offer the president may, in the end, tie his hands.

Conclusion

Current and future UCAVs offer the politician many advantages. But the extent to which these advantages appeal to them will largely depend on the type and the nature of conflict. UCAVs may not offer the politician distinct advantages over current systems in wars broadly supported by domestic and international audiences. This changes, however, when conflicts erupt that are difficult to justify domestically and internationally. Reduced casualties, smaller footprint, and modern command and control technology will, in the end, combine to offer the politician a seemingly high degree of control over situations involving conflict. This combination may be the most appealing aspect of these new weapons, but one that may carry with it hidden dangers.

Chapter 2

Moral Implications of UCAVs

This chapter addresses moral concerns in two ways. The first section examines UCAVs in light of currently accepted principles, rules, or norms that govern the weapons of war arguing that UCAVs, by themselves, do not violate any of the above standards. Next, the chapter examines UCAVs against the more ambiguous standards of chivalry and fairness, arguing that neither standard is relevant to the discussion. Ultimately, the purpose of this chapter is to provide strategists greater clarity on the moral issues surrounding UCAV employment. Doing so allows them to more effectively address, both objectively and subjectively, the moral implications of this new technology.

Principles and Rules Governing Weapons in War

The most commonly accepted standards that most nations use to evaluate the morality and legality of new technology are Just War Theory and the Laws of Armed Conflict (LOAC).

Just War Theory

Early religious scholars and philosophers spent a great deal of time pondering the morality of war. They determined that although war is bad, it is sometimes necessary. And because it is sometimes necessary, there should at the very least be rules that the parties should follow to prevent war from becoming an end in itself resulting in societal chaos.[41] Their thoughts, collectively known as Just War theory, consist of two distinct parts. The first part, *jus ad bellum*, provides a list of nine principles that states should consider before engaging in conflict. The second part, *jus in bello*, provides guidelines governing the conduct of states once they are engaged in war.[42] Both parts of the theory are important to the issue of the morality of UCAVs, but it is the latter that is the most relevant to the discussion at hand.

---

[41] Bruno Coppieters and Nick Fotion, "Introduction," in *Moral Constraints on War: Principles and Causes,* ed Bruno Coppieters et al. (Lanham, MD: Lexington Books, 2002), 11.

[42] Michael Walzer, *Just and Unjust Wars: A Moral Argument with Historical Illustrations,* (New York, NY: Basic Books, 2000), 21.

Jus in bello

The *jus in bello* principles of discrimination and proportionality provide a framework for the strategist to use when judging the morality of weapons and methods in war. Discrimination revolves around non-combatant immunity. "In the broadest sense, the principle maintains that warring parties have an obligation to discriminate between appropriate and inappropriate targets of destruction, a distinction based on the nature of the targets themselves."[43] Appropriate targets are recognized combatants and the military infrastructure that enables them to continue fighting, whereas those individuals not involved in the war and designated as non-combatants by treaty and convention are off limits.

Proportionality is a principle found in both *jus ad bellum* and *jus in bello*. When used in *jus ad bellum* it asks a state to determine whether the "anticipated moral cost of fighting [a] war [is] in line with the moral benefits."[44] Simplified, is the objective of the war moral, and if so, is the objective worth fighting and killing for? Conversely, proportionality in *jus in bello* asks combatants themselves to determine whether the "degree of violence employed in pursuing military objectives [is] proportional to the significance of the military objective."[45]

Theorists admit the possibility of civilian or non-combatant deaths from the inherent violence associated with war. However, they desire to limit and minimize this destruction as much as practical. Just War Theory requires military commanders to evaluate their wartime decisions, taking into account the possibility that their means and methods of fighting war might contribute to civilian destruction and suffering. If their actions have the possibility of producing excessive civilian casualties, then the commanders are morally obligated to seek alternative means or methods to accomplish their objective.

Discrimination, Proportionality and UCAVs

---

[43] Anthony Hartle, "Discrimination," in *Moral Constraints on War: Principles and Causes,* ed Bruno Coppieters et al. (Lanham, MD: Lexington Books, 2002), 141.
[44] Guy Van Damme and Nick Fotion, "Proportionality," in *Moral Constraints on War: Principles and Causes,* ed Bruno Coppieters et al. (Lanham, MD: Lexington Books, 2002), 129.
[45] Hartle, 172.

More often than not, whether discussing soldiers or UCAVs, it is the way in which a weapon is used and not the actual weapon that determines the morality of its use. For example, modern soldiers are basically just very intelligent weapons delivery platforms that carry a host of different types of legal weapons. Yet they can use them in both moral and immoral ways. Armed with a machine gun, a soldier has the capability to shoot either lawful combatants or innocent civilians. If he targets the former and subsequently kills the latter in the process he has acted immorally by failing to adhere to both *jus in bello* principles.

The same holds true whether discussing an F-16 or a UCAV in that they are both merely weapons delivery platforms. Both can be equipped with various weapons and both can be used in ways that violate proportionality and discrimination principles. Simply removing the pilot from the cockpit does not in itself make an aircraft immoral, as long as UCAV operators have the ability to control the actions of their aircraft.

One of the few moral arguments against UCAVs to recently surface in the media dealt with a Predator strike in Yemen. In November 2002, the CIA used a Predator equipped with a Hellfire to engage and kill a top al Qaeda leader believed to have been behind the attack on the USS Cole in 2000. The Swedish Foreign Minister claimed it was "a summary execution that violate[ed] human rights" because the Predator operator was in effect acting as judge, jury, and executioner. However, others disagreed with that assertion, claiming that because the United States had declared war on al Qaeda, the strike was a legal military strike, not an assassination.[46] Though it was a CIA and not a military strike, the case provides some insight into the discussion about methods, rather than the actual weapon, determining the morality of a weapon. The morality of the UCAV weapons delivery platform was not debated: if the strike had been carried out by an F-16, the issue of 'judge, jury, and executioner' would not have been altered. What was up for debate, though, was the morality of how the platform was used.

Opponents of UCAVs might argue, however, that UCAVs in and of themselves are disproportional. A statement like this is likely made out of ignorance of what the principle means as it is often misinterpreted. Proportionality in no way implies that

---

[46] Vince Crawley and Amy Svitak, "UAV Strike Raises Moral Questions," *Air Force Times,* November 18, 2002, 16.

casualties on both sides need to be in proportion to each other in order for the *jus in bello* principle to remain unviolated. For instance, just because the United States overwhelmingly defeated the Iraqi military in Desert Storm with far fewer casualties than the Iraqi side does not mean that the forces or the equipment the United States used was out of proportion to the objectives sought.[47]

UCAVs, in fact, actually increase the commander's ability to adhere to proportionality concerns. If the commander is morally obligated to use the minimum force necessary in achieving his objective against enemy combatants in an attempt to reduce casualties on the enemy's side, he is also morally obligated to do the same for his forces. History is replete with examples of the military leader using technology to increase the safety and security of his forces to enable them to fight another day. Removing the pilot, therefore, allows the commander to do this more efficiently because it reduces the chance of unnecessary casualties. If the intent of just war theory is to "maximize good and minimize evil" then decreasing casualties, regardless of on which side they occur, is a moral requirement.[48]

UCAVs may actually increase the operator's ability to adhere to *jus in bello's* discrimination requirement. First, new sensor technology available on today's UCAVs provides much better acuity than that of the unaided eye. Moreover, UCAVs have multiple sensor options to deal with a wide range of meteorological and time of day conditions. Modern sensors have the ability to switch from infrared to electro optical or use radar for targeting. Granted, some of that same technology is available in manned cockpits, but the ability to interpret this information in the clean, risk free environment of a control trailer 5000 miles away is extremely valuable during the chaos of war.

Second, the increase in distance from conflict provides a reduced stress environment that allows the operator of a UCAV to focus on the task at hand. When a pilot or a crew is faced with the task of engaging targets in a manned aircraft, they have to deal with a host of additional stressors and concerns. These range from concerns over maintaining formation integrity while dealing with an enemy's air defense network, to receiving target change instructions that greatly complicate the mission. These stressors

---

[47] Hartle, 172.
[48] Van Damme and Fotion, 130.

can lead to mistakes that may increase collateral damage and fratricide occurrences. UCAV operators do not operate in an entirely stress free world; however, the total overall level of stress is less than those in manned cockpits. Operators are still encumbered by the stress of adhering to legalistic rules of engagement, maintaining basic aircraft control, avoiding enemy fire, and remaining within the confines of restricted airspace, but this stress is tempered by the fact that you are not likely to die.[49]

Finally, contrary to popular opinion among flyers, overall situational awareness may actually increase for UCAV operators over their manned counterparts. Increased situational awareness helps reduce incidences of collateral damage and fratricide. Lt Col Steve Luxion, commander of the first weaponized Predator squadron, had the following to say about Predator employment and situational awareness during operations in OEF and OIF:

> "My situational awareness was better [as a Predator operator] than it ever was when I was in the cockpit. I'm updated immediately or near real time or as fast as information can flow and get to me by off-board systems such as Link 16, blue force tracker, or other off-board sources. The picture of everything I had going on around me was absolutely fantastic. If asked to change targets, I could have a picture [in front of me] within 30 seconds to a minute and have someone briefing me [about the new target]. All this helped to increase my SA before I even got there."[50]

Luxion claims that operations centers created to support UCAV operations provide synergies not available to the pilot of a manned aircraft. He calls these centers 'fusion cells' and in explaining their use, likens them to NASA control facilities. "It's almost like being a [space] shuttle pilot, in that you have this big ground based system of people and engineers all supporting that pilot in their mission. These cells can help the operator make rules of engagement calls."[51] Although the pilot is still ultimately responsible for weapons release decisions, he has help from many sources and people who would not necessarily be there to help him if he were in the cockpit. The situational awareness he receives from the fusion of information available helps him make more

---

[49] Lt Col Gary Fabricus, Squadron Commander of one of the first two Weaponized Predator Squadrons, interviewed by author 14 January 2005.
[50] Lt Col Steve Luxion, Squadron Commander of one of the first two Weaponized Predator Squadrons, interviewed by author 12 January 2005.
[51] Luxion interview, interviewed by the author 12 January.

informed decisions when issues of discrimination or collateral damage concerns arise. While acknowledging that fog and friction still exist, he does claim that they "seem somewhat less than if you were in the cockpit."[52]

Treaties, Conventions, and LOAC

As nation-states evolved, many came to accept the Just War standards. Though the level of adherence by states varied at times, the international community codified these standards and norms into international treaties and conventions such as the Hague Treaties and the Geneva Conventions. The international community references these treaties when making moral and legal determinations about the conduct of war. But in some instances, these documents and their legalistic prose are difficult to understand, leaving room for interpretation. To simplify the commander's responsibility, military legal experts have interpreted the conventions and treaties and consolidated them into what is commonly referred to as the Laws of Armed Conflict, or LOAC. Military lawyers consider the LOAC as the broad-based rules that define how to fight war, rules that military members have a legal duty to observe.[53] These are the same rules that dictate how the commander employs weapons in war.

Because of the foundational similarity between the two standards, commanders can use UCAVs in ways that abide by LOAC principles just as they can with regards to Just War principles. There are five basic principles in the LOAC: military necessity, distinction, proportionality, chivalry, and humanity. Military necessity requires that attacks be limited to military objectives. Distinction requires the military to distinguish between military objectives and civilian objects. Proportionality requires that the military commander take into consideration the extent of civilian destruction and probable casualties that will result and look for ways to avoid or minimize such casualties and destruction – civilian losses must be proportionate to the military advantages sought.[54]

---

[52] Luxion interview, interviewed by the author 12 January.
[53] Michael W. Goldman, ed., *The Military Commander and the Law,* (Maxwell AFB, AL: Air Force Judge Advocate General Press, 2004), 558.
[54] Goldman, 548-553.

The intent of these first three is the same as the *jus in bellum* principles of proportionality and discrimination. Since military commanders can use UCAV in ways that adhere to *jus in bello* principles, it stands that they can also use them in accordance with similar standards outlined in the LOAC.

In the LOAC, the principle of chivalry requires the military commander to wage war in accord with well-recognized formalities and courtesies. Specifically, it forbids the commander from taking unfair advantage of the enemy through the misuse of internationally recognized symbols such as the white flag of surrender or the Red Cross or Red Crescent. For example, it is illegal for a commander to launch an attack under the guise of a white flag of surrender. It would be equally unlawful for a commander to use a Red Cross vehicle to move troops around the battlefield.[55] Though not in violation of the LOAC standard of chivalry, the use of UCAVs does challenge other long-held beliefs about chivalry not covered under this standard, a subject addressed later in this chapter.

The fifth principle, humanity, deals directly with moral and immoral weapons and is the most relevant in terms of making determinations about the morality of the UCAV. Because weapons of warfare are ever evolving, this principle provides broad guidelines for the commander to use in making determinations of lawful and unlawful weapons. Specifically, it forbids the commander from using three types of weapons: those employing poison (chemical or biological weapons); those arms, projectiles, or material that causes unnecessary suffering (hollow point bullets); and those that treacherously kill or wound individuals.

The treacherous provision refers the class of weapons that have an inherent ability to kill indiscriminately. Therefore, biological, bacteriological, and chemical weapons are all immoral according to this provision. Furthermore, this provision outlaws weapons that cannot be controlled.[56] UCAVs and other remotely piloted vehicles are fully controllable from takeoff to landing and therefore do not violate this provision. Even in the event that connectivity is lost between the operator and the aircraft, they still do not violate this provision. In this case, the UCAV continues on a pre-programmed flight path

---

[55] Goldman, 548-553.
[56] The controllability provision is not meant to exclude weapons that are uncontrollable after launch, release, or firing. Bullets, ballistic missiles, and dumb bombs, for instance, are designed and employed with tested ballistic profiles, and are therefore not considered uncontrollable by LOAC. Goldman, 548-553.

until communication links are reestablished. The software inhibits weapons arming and release commands until the operator gains full control of the aircraft.

Some critics question the legality of the UCAV on grounds that it violates the 1988 Intermediate-range Nuclear Forces (INF) Treaty signed by the Soviet Union and the United States. The INF treaty prohibits either country from deploying ground-launched cruise missiles with ranges between 500 and 5,500 kilometers. The types of missiles addressed in the treaty are built for one-time use and fly a one-way profile without the ability to return to base. Current and future UCAVs, however, are built for reuse and therefore do not violate the INF.[57]

In sum, UCAVs are really no different than other types of aircraft used in war with regards to the internationally-recognized norms of Just War theory and LOAC criteria. Like most other weapons, it is the way that the military commander uses them that determines their morality. In fact, combining the UCAV with near real-time intelligence and precision engagement capabilities may leave the U.S., in the words of one analyst, "well positioned to adhere to just war principles."[58]

Moral Ambiguities

Assuming that the commander will employ UCAVs in accordance with the standards espoused by the LOAC and Just War theory, are there other reasons for questioning their morality?

As humans we become used to standard ways of doing things – business, law, war – and when something new comes along, our first reaction is to determine if it is right or wrong, just or unjust, moral or immoral. Is there some concern that using UCAVs to fight wars is somehow wrong because it just does not feel right? Usually when something is legal yet does not feel right it usually means that it is "morally uncertain."[59] In an effort to deal with this moral uncertainty, professions have often established ethical standards as guidelines for behavior.

---

[57] Anthony J. Lazarski, "Legal Implications of Uninhabited Combat Aerial Vehicles," *Aerospace Power Journal*, Summer 2002, 79-80.
[58] Andrew Bacevich, "Morality and High Technology," *National Interest,* Fall 1996, 6.
[59] Bud Cameron, "When Robots Kill," (masters thesis, Canadian Forces College), 14.

While morality concerns "the quality of being in accord with standards of right or good conduct," ethics address "the rules or standards governing the conduct of the members of a profession."[60] Professions, be they medical, legal, or military, follow established rules and guidelines – ethics – created to continue the legitimacy of that profession. "Ethics are enablers … personal, social, or military [that] allow us to interact without needless viciousness and without generalized violence to the soul, the body, or society."[61]

As warfare has evolved, the military has carried with it general guidelines or a professional ethic of how it thinks war should be fought. Some of these guidelines are based in the principles, rules, and norms that this chapter has already addressed. But others are wrapped up in long held beliefs of how war should be waged, beliefs that vary from culture to culture as well as from one time period to another. Western military tradition and habit – ethics – are wrapped up in notions of chivalry and fairness in combat. These notions, however, are irrelevant to the issue at hand and therefore should not be used to limit the advance and employment of UCAVs.

Chivalry

Some of the moral uncertainty associated with UCAVs likely stems from the notion that employing UCAVs does not seem chivalrous. The idea of an ill-trained Iraqi soldier fighting against a modern Hellfire-equipped Predator or a Mig-29 pilot against an air-to-air UCAV piloted 5000 miles away just does not seem chivalrous in any sense. But what is this sense of chivalry based on, and does it still apply to modern day warfare?

The concept of chivalry has its roots in the Middle Ages. John Lynn agrues that, idealistically, "[chivalry] is an elaborate and elegant discourse on war, carrying with it conceptions of combat and of proper conduct toward leaders, fellows, foes, and non-combatants."[62] Working in the service of their lords, knights met each other on the battlefield or during jousts to settle differences between the nobles and even the knights themselves. Implicit in the chivalric ideal of fighting was the notion that the knights should face each other on the battlefield on equal terms. For example, if a knight fell

[60] *New American Heritage Dictionary,* (Boston, MA: Houghton Mifflin Company, 1982)
[61] Ralph Peters, "A Revolution in Military Ethics," *Parameters,* Summer 1996, 1.
[62] John A. Lynn, *Battle: A History of Combat and Culture,* (Cambridge, MA: Westview Press, 2003), 78.

from his horse during the course of battle, the opposing warrior was to allow him to re-mount in order to continue the fight in a fair and even manner.[63]

In practice, chivalry varied greatly from its idealistic foundation. Rules of warfare that the medieval knights followed in the service of the nobility were put in place to make war safer only for the "aristocratic elite."[64] Rather than kill the opposing side's nobles and knights, medieval warriors sought monies from the ransom of captured subjects; profit was the overriding concern. According to Christopher Coker, chivalry was only concerned with how to treat the fellow elite or aristocratic warriors and had little to do with the rest of society.[65]

If a chivalric code did anything, it perhaps made war more inhumane, particularly for those who were on the sidelines. While awaiting the payment of ransom, soldiers would attack and pillage towns and in the process, killing or starving innocent townspeople. Had the knights simply killed the opposing nobility or knight, the conflict would have been resolved quickly, thereby allowing innocent townspeople to get back to their lives. Thus acts considered chivalric could also be characterized as ultimately increasing the pain and suffering of non-combatants and clearly immoral by Just War standards of maximizing good and minimizing evil. Therefore, as Coker suggests, "we should not be fooled by the 'humane' code of chivalry."[66]

Changing world events, though, undercut the foundations of chivalry. The Treaty of Westphalia in 1648 ended the Thirty Years War in Europe and the conflict between the Catholic and Protestant forces, giving birth to the system of nation-states. Subsequently, nations fought wars primarily for reasons of state, with religion taking a back seat to state interests.[67] Moreover, after the French and Industrial Revolutions, states had at their disposal the men and equipment needed to place large armies into the field to protect the homeland. These armies were expensive to train and maintain and so it was in a state's best interest to quickly fight decisive battles and preserve forces for future wars.[68] Therefore, states continually looked for ways to protect and reduce the risk to their

---

[63] Lynn, 78.

[64] Lynn, 77.

[65] Christopher Coker, *Humane Warfare*, (New York, NY: Routledge Press, 2001), 122-124.

[66] Coker, *Humane Warfare*, 122.

[67] Sir John Hackett, *The Profession of Arms,* (New York, NY: Macmillan Publishing Company, 1983), 75.

[68] Christopher Coker, *Waging War without Warriors: The Changing Culture of Military Conflict*, (Boulder, CO: Lynne Rienner Publishers, 2002), 53-55.

soldiers through constantly improving technological means. Thus, the concept of chivalry, meant to reduce the cost of war and line the pockets of the knights and nobility, was overtaken by rules of war that sought to minimize the ultimate cost of war for the state in both blood and treasure.

Chivalry and morality, thus, are not inextricably linked. Just because a weapon or method of conducting warfare does not appeal to our chivalric sense (outdated as it may be) does not mean it is immoral. Therefore, tying chivalry to morality in an argument against the use of UCAVs is unfounded and akin to arguing against technological advancement by, as Eric Cohen has written, "romanticizing about lost worlds and pretending that societies without running water or modern medicine are more authentic [and thus better] than our own."[69]

Fairness

Perhaps the moral uncertainty leading some to question the employment of UCAVs centers on the issue of fairness. On the surface, it hardly seems fair for an individual to kill another without risking his or her own life in the process. And because society equates the issue of fairness with justice and justice with morality, killing in this manner could be characterized as immoral.[70] Yet such a claim would be based on confusion between the idea of a fair fight and an even fight. Society generally resolves issues of fairness by instituting rules and principles that governs the conduct of individuals in certain endeavors. As long as individuals abide by these rules and norms they are free to seek ways to increase their advantage. Simply acting in ways that maximize one's advantage does not necessarily make an act unfair. This is different from the idea of an even fight. In an even fight, both sides are equally equipped with similar instruments or means, giving neither side a clear advantage.

Sports coaches do not look for an even match when developing game plans against opposing teams. For example, football coaches spend endless hours studying film of opposing teams looking for weaknesses they can exploit. They devise trick plays to take advantages of these weaknesses. As long as the trick plays conform to the NFL

---

[69] Eric Cohen, "The New Politics of Technology," *The New Atlantis: A Journal of Technology and Society*, Spring 2003, on-line at www.thenewatlantis.com/archive/1/editorialprint.htm, 10 October 2004, 4.
[70] Fotion and Coppieters, 15.

rules concerning the play of the game, the coach is free to use them at his discretion. If the plays do not conform to the rules, then any gains from their use are deemed unfair and subsequently illegal.

Similarly, military commanders that employ UCAVs are also looking to increase their advantage over the opponent. In choosing the weapons of war, commanders only have to abide by the principles, rules, and norms that govern warfare. When operating within those bounds, they are free to seek ways to increase their advantage just like the football coach does when making and executing his game plan. To expect commanders to handicap themselves in order to make a fight more even is unrealistic. "Charity in warfare is either a non sequitur of an invitation to defeat," and no commander worth his salt wants an even fight.[71] The United States has spent a great deal of money developing tools for warfare to do just that, and the UCAV just happens to be the latest development in that quest. If you categorize the UCAV as unfair, then you also have to make a similar assertion about other means of warfare that seek to increase the distance between the shooter and the target such as airplanes, submarines, or ICBMs.

Conclusion

Many of the concepts against which UCAVs are judged have a temporal dimension to them. It is doubtful that the chivalry of medieval times existed 2000 years ago in the days of the Alexander the Great. Chivalry was useful only at a certain time and place under special circumstances. In today's environment, chivalry may not be the best standard for judging the acceptability of new technologies. Likewise, the laws of war are also temporal but in a different way. While the football analogy is simple, its use leaves out one important distinction. In war, unlike football, the rules can change or even be discarded. This is especially true for wars of national survival, a case Walzer describes as a "supreme emergency."[72]

Due to the fast changing American way of war, strategists must continually evaluate "the manner in which the U.S. fights" to ensure that it "reflects the values of just

---

[71] Hartle, *Moral Constraints on War*, 172.
[72] Walzer, *Just and Unjust Wars*, 251-262.

war tradition, values that are consistent with the laws of war and with American ideals."[73] Therefore, before passing judgment on the morality of a new technology, strategists must first measure technology objectively to determine if it, or its use, violates any generally accepted rules and principles. Next, they should ensure that their judgment is not clouded by yearnings for the way it use to be. Just because a new technology does not nicely fit into the moral container society is used to seeing and using does not mean it is unfit for use. Moreover, if a new technology's arrival prompts some to question the legality or morality of its use, it is unwise to ignore their concerns. Most likely there is something worth paying attention to in their first, instinctive response. The strategist should attempt to get to the core of this instinct so that they can form a coherent argument to address the concerns either preemptively or after the fact. Doing so allows them to counsel and advise those who ultimately make the decision to use the new technology. If strategists know the issues surrounding new technology, they will be better able to justify its use to the public.

---

[73] Daniel E. Soller, "Operational Ethics: Just War and Implications for Contemporary American Warfare," (masters thesis, School of Advanced Military Studies, The Army Command and General Staff College), abstract.

Chapter 3

Social Implications of UCAVs

One possible implication of the employment of UCAVs is that a professional military ethos will no longer matter in the age of the remote-control battlefield. This is based on the premise that anyone can perform combat duties once reserved only for the military if risk and danger are removed. But does remote-control war really change the underlying assumptions that have been the basis for the military ethos in the past? Or do society and the military tend to confuse the idea of a professional military ethos with a warrior ethos?

The Military Profession

Noted sociologist Samuel Huntington argues that in order for a vocation to be considered a profession, it must have three characteristics. First, the members of a profession have to have a certain level of *expertise* in order to conduct their business. This expertise is not the kind associated with building a house or repairing a car. It requires a greater level of intellectual thought and knowledge associated with past accomplishments of complex and demanding tasks. The second characteristic revolves around the idea of *responsibility*. The members of a profession perform functions that are essential to society at large. They feel called to serve in this capacity and do not take their profession lightly. "The responsibility to serve and devotion to [their] skill furnish the professional motive." Finally, the members of a profession exhibit a sense of *corporateness*. Corporateness is largely a byproduct of lengthy training and common hardships. It is further fostered by a unique sense of self that comes from knowing that a particular job cannot be performed by the masses. These characteristics are incorporated into a written code of ethics that guides the profession.[74]

Huntington argues that the military is also a profession much like the medical and legal professions. The military member's *expertise,* however, resides in his ability to manage violence. This is a skill that is the central distinguishing feature that separates

---

[74] Samuel P. Huntington, *The Soldier and the State: The Theory and Politics of Civil-Military Relations* (Cambridge, MA: The Belknap Press of Harvard University Press, 1957), 8-10.

the military profession from the others in society. Unlike the doctor who vows to do no harm, the military member has to be prepared and is sanctioned by the state to do just the opposite – to kill. The act of killing is in direct opposition to societal norms, further distinguishing it from other professions. With this expertise, comes a great deal of *responsibility*. The military member is given the responsibility to defend and serve his country, to protect it from enemies both foreign and domestic. Moreover, he has the power to endanger the lives of his men as well as the lives of the enemy, a responsibility that cannot be taken lightly. Huntington's *corporateness* is the military's esprit de corps. Shared hardships that start with basic training and continue with hardship duty in the field foster a sense of camaraderie. Additionally, they know they have a unique skill set that separates them from the rest of society and this acts to draw them to each other. Only other members of their profession understand the true nature of their profession, especially since it requires them to violate societal norms against killing.

All professions have a code of ethics and the military is no different in this regard. The military's professional code of ethics guides individuals in their duties, providing boundaries for them to operate within. This is especially important given the dynamic and frequently changing character of war.[75] Selfless service, integrity, honor, and duty are just a few words that describe this ethic. The ethic has several purposes. First, it bridges the gap with society, reinforcing the idea that the soldier serves the state.[76] Second, it guides the individual in carrying out his duties in war, reminding him that what he does, whether it involves killing or not, has great consequence for him, his fellow service-members, and the country he or she serves. The ethic applies equally to all military members regardless of their duties. Each one of these individuals is authorized by the state to kill the enemy. Furthermore, each one is asked to sacrifice their life if necessary. It does not matter that some of the individuals are less likely to experience direct combat action than others. The important element is that they can be called on if necessary to do so.

Many in society and the military in general refer to the *professional military* code as the *warrior* code. The warrior code serves a distinct purpose in that it guides the

---

[75] Huntington, 11-18.
[76] Phillip M. Flammer, "Conflicting Loyalties and the American Military Ethic," in *War Morality, and the Military Profession*, ed. Malham M. Wakin (Boulder, CO: Westview Press, 1979), 164.

warrior in battle. Shannon French, in *Code of the Warrior*, asserts that the code defines "how he should interact with his own warrior comrades … other members of his society … and the people he conquers." If a soldier adheres to the warrior code, he will not flee in the face of danger. He will sacrifice his life for the greater good, whether it is his country or his fellow warrior. He will also treat his enemy with respect. Additionally, the code acts to restrain the warrior. "It sets boundaries on his behavior … distinguishing honorable from shameful acts" in an attempt to "protect the warrior himself (or herself) from serious psychological damage."[77]

Warriors who follow the code are said to embody a warrior ethos. The military services have each attempted to identify the traits that a warrior must possess in order to have a warrior ethos. While each service has its own definition of what the warrior ethos is, they all revolve around a list of specific traits such as sacrifice, courage, bravery, aggressiveness and discipline. The warrior ethos can be best described as "the character, values, behaviors, and attributes developed within groups of warriors over centuries of armed combat which are essential to closing with and destroying the enemy."[78]

It is important to remember that, though some may view the professional military code or ethos and the warrior code or ethos as separate and distinct, they are not. They are one in the same in that they provide guidelines and standards of conduct for the members of the military who are asked to kill regardless of distance from the enemy when the killing takes place. The state has sanctioned and authorized the military member to kill for purposes of the state. This is what separates him from the rest of society. The confusion exists because people tend to get caught up in the word *warrior*. If a duty does not involve danger or risk, it must not be a duty required of a warrior. They forget that it is the end result of the warrior's duty – killing – that provides the reason for the code or ethos. This is the most relevant aspect of the code.

The Air Force and the Warrior Ethos

The Air Force constantly emphasizes the idea of the warrior ethos, desiring its members to identify with the ethos and embody those traits that lead to success in battle.

---

[77] Shannon E. French, *The Code of the Warrior: Exploring Warrior Values Past and Present* (Lanam, MD: Rowman & Littlefield Publishers, Inc., 2003), 5-6.
[78] David W. Buckingham, "The Warrior Ethos" (master's thesis, Naval War College, 1999), 4.

Senior leaders in the Air Force never fail to mention and emphasize the importance of a warrior ethos to their audiences when discussing the responsibility of serving the country. Yet they tend to view it through a lens that focuses on the aspect of closing with the enemy to kill him rather than the actual act of killing itself. In a speech to graduating Air Force Academy cadets, Former Secretary of the Air Force, James Roche, stated that "the warrior ethos demands such high standards of accountability, maturity, and the *acceptance of personal danger and individual risk.* "(emphasis added)[79] The last part of this quote amplifies the idea that Air Force leaders themselves equate ethos with risk – the risk involved in closing with and killing the enemy. But what happens when the person or a group of people in the organization does not or cannot tie his or her job directly back to the warrior ethos that these leaders emphasize? What if their duty does not include personal danger or individual risk? Does this cause them to look disparagingly on duties that do not directly fit those associated with the warrior in the classical sense? Does this tend to alienate them from the group?

Emphasizing the closing with the enemy aspect of the ethos over the actual act of killing itself may be sending the wrong message to certain groups or individuals in the military. Individuals in certain specialties realize that their duties do not require them to close with the enemy. They realize there is a misfit between the notion of the warrior in the classical sense and the requirements of their job in modern warfare. Continually trying to force fit the classical warrior ethos on to all duties does nothing more than alienate those who cannot readily tie their job back to historic ideals. This has happened in the Air Force with unmanned aircraft operations in general. Pilots generally have had a disparaging view of UAV operations because the operational aspects of UAVs do not fit what they have come to expect or associate with warfighting. Furthermore, the Air Force as an organization itself has in the past failed to embrace this technology fully, relegating those that fly them to second class status.

The Air Force Personnel Center has historically filled UAV positions with pilots or navigators holding a civilian commercial aviation license due to Federal Aviation Administration regulations. There was no requirement for the operator to have

---

[79] Secretary of the Air Force, James G. Roche, "The Centennial Airmen – A New Generation of Air and Space Leaders," commencement address, Air Force Academy, CO, 28 May 2003.

experience in a specific type of airframe, although 80% of a typical squadron came from the fighter and bomber community with the remaining 20% coming from airlift and other communities. Many, however, did not volunteer for the two year assignment and were initially displeased when notified.[80]

There were many reasons for their disappointment. To begin with, flying a Predator is much different than flying fast and sleek airplanes that society associates with military aviation. Pilots were being asked to go from "flying an F-15 at 30,000 feet going Mach 1.5 [to] sitting in a trailer watching a TV screen," controlling the 115 mile per hour Predator with controls akin to a modern day video game. Predator operations were not very glamorous for individuals who "felt like they [had] been trained for something much grander and more difficult."[81]

Moreover, the Air Force has not always treated a tour in the Predator like a normal flying assignment. Up until 2002, the Air Force did not count Predator flight time toward a pilot's total gate time.[82] Gate time credit is an accounting measure that the Air Force uses to determine how long a pilot will continue to receive flight pay.

The culture of the Air Force flying community itself added to feelings of inadequacy. It is a culture where operators identify themselves with their respective airframes more so than their occupation. If you ask an aviator what he does in the Air Force, he is likely to answer with "I'm a bomber pilot" or "I'm a Viper (F-16) pilot." Some even consider themselves pilots first and Air Force officers second.[83] But ask a predator pilot what he flies and he's likely to say "I'm a former Viper (Eagle, C-5, B-1) pilot, but I fly Predators now."

This culture places a wall between the pilot in the air and the one on the ground. It comes from the feeling that the pilot in the sky is the only true warrior because he "can get shot down or killed" while closing with the enemy, whereas the pilot in the control

---

[80] The Air Force recently decided to draw all future weaponized Predator pilots exclusively from the fighter and bomber community looking to draw on their previous experience in weapons employment. Frank Colucci, "Air Force Refines Training Programs for UAV Operators," *National Defense*, May 2004, 37-39.
[81] Ann Marie Squeo, "Top Guns Grounded: Pilots Fume at Duty on Unmanned Craft --- Fliers Used to F-16s Now Sit in Windowless Cubicles; Shot Down at the Arcade" *Wall Street Journal*, April 29, 2002.
[82] James G. Roche, Secretary of the Air Force, memorandum for ALMAJCOM-DRU/CC. subject: Air Force Policy Regarding OFDA Credit for UAV Crews, 25 April, 2002.
[83] Carl H. Builder, *The Masks of War: American Military Styles in Strategy and Analysis* (Baltimore, MD: The RAND Corporation, The Johns Hopkins University Press, 1989), 23.

center "is sitting back in an air conditioned building flying from Las Vegas."[84] It does not matter that the end result from both of their activities are the same. In the current Air Force culture, distance matters.

This separation does not stop at the door of the UAV squadron. A wall also exists between the tactical aviators, those who have flown fighters or bombers, and the non tactical ones, those who have flown airlift and tanker aircraft. According to one commander, "This diversity created a lot of conflict within the squadron."[85]

This conflict is just an extension of the culture within a culture that exists in the Air Force aviation community in general. Carl Builder calls these intraservice distinctions. He argues that they tend to drive a wedge between members of those sub-communities, ultimately resulting in a less than optimum organization.[86] Perhaps this distinction within Predator squadrons is exacerbated by the lack of respect that UAVs have as a weapon system. For many, a UAV is just not sexy or noble. Some in the community equate it to riding a moped: yes, a moped is a useful and efficient means of transportation, but you do not want any of your friends to see you riding one.[87]

In the past, this organizational culture made it difficult for Predator pilots to even consider themselves pilots, much less warriors. For all these reasons operators did not seem proud of what they did, perhaps failing to find honor or courage in their duties. They failed to feel a sense of mission or duty what the Air Force asked them to do. They found themselves marking time, waiting for the two year tour to end so they could go back and do the Air Force mission that they had signed up for. Operators did not believe they were contributing to the warrior mission of closing with and killing the enemy many had signed up to do. They did not feel respected by their peers or by the Air Force institution itself. Even the Secretary of the Air Force acknowledged this after a close-up look at Predator operations. Based on complaints he heard from some of the operators,

---

[84] Lt Col Gary Fabricus, Squadron Commander of one of the first two Weaponized Predator Squadrons, interviewed by author 14 January 2005.
[85] Lt Col Gary Fabricus, interviewed by author 14 Jan 2005.
[86] Builder, *Masks of War*, 24-27.
[87] Lt Col Steve Luxion, Squadron Commander of one of the first two Weaponized Predator Squadrons, interviewed by author 12 January 2005.

Secretary Roche said that "The kids [operators] were right. Their feelings were somehow being shunted."[88]

UCAVs and the Warrior Culture

These feelings of inadequacy or alienation were largely the result of a culture that has emphasized the aspect of closing with the enemy in order to kill him as being the most important part of the ethos when it is instead the actual act of killing itself that is most relevant.

Admittedly, UCAVs do not fit neatly within the traditional warrior culture. The operating environment of the UCAV operator is much different than that of the traditional ground or air combatant. Their workspace is an air-conditioned operations trailer. Additionally, operators can take breaks when necessary, handing off their responsibilities to other operators when they get tired or hungry, something that the battlefield or battlespace combatant cannot do. Furthermore, in performing their duties they encounter little if any risk from direct combat. UCAV operations, therefore, are relatively benign and risk free when compared to the battlefield combatant. Predator operator Lt Col Kurt Scheible describes it this way: "When I'm back in Nellis [Air Force Base, located in Las Vegas, Nevada] I can fly a mission over Iraq with the Predator, and then go home and take my children to a ball game."[89]

It is important to remember, however, that though UCAV operators do not fit neatly into the warrior culture, ultimately they are performing the same duty as the warrior on the battlefield or in the skies. True, they do not "close with" the enemy, but the UCAV operator is sanctioned by the state to engage with and destroy him. He is expected to do so while adhering to the same rules and laws that guide the military professional who closes with and kills the enemy. His duties easily fit under Huntington's description of the military professional as a manager of violence.

Furthermore, removing risk and danger does not obviate the need for the operator to have a professional military ethos. He or she still needs an ethos that guides and

---

[88] Ann Marie Squeo, "Top Guns Grounded: Pilots Fume at Duty on Unmanned Craft --- Fliers Used to F-16s Now Sit in Windowless Cubicles; Shot Down at the Arcade" *Wall Street Journal*, April 29, 2002.
[89] Stephen Grey, "Pilotless Strikes on Iraq by RAF," *London Sunday Times*, 3 October 2004. Article discusses joint UK/US Predator operations at Nellis Air Force Base.

constrains behavior regardless of distance from the battlefield. The larger point is that even though all members in the military do not fit the stereotypical warrior ideal, they are in fact warriors by the intent of the word and therefore subject to a professional military code of ethics. We should not get caught up in the rhetoric over strict definitions of what a warrior is for it is the intent that matters most. This is especially important for the Air Force, a service that has constantly sought to increase the distance between the operator and the enemy.

Conclusion

This is not a call for the Air Force to stop or reduce its emphasis on the warrior ethos. This is instead a call for the Air Force to emphasize the responsibilities inherent in the sanction to kill that underlies the warrior ethos. Service leaders should acknowledge that some airmen will close with the enemy while others will not. However, the end result will be the same. The ethos exists to provide guidelines to the airmen in carrying out his or her duties, duties that are unique and distinctive from the rest of society. Changing the emphasis from one centered on closing with the enemy to one centered on the actual act of killing is especially important as technology continues to further remove the warrior from the battlefield. If, however, the Air Force fails to account for the changing nature of warfare in general and its mission in particular, it may end up with a subset of the profession that merely views their job as an occupation rather than a profession. Killing is too important a task to be relegated to occupational status.

Chapter 4
Unintended Consequences

Edward Tenner, in his book *Why Things Bite Back,* gives numerous examples of how man has tried to use technology to benefit himself and make life simpler and safer, yet in doing so has created other unexpected or unintended problems in the process – even to the point of obviating the need for the technology in the first place. He claims that "[w]henever we try to take advantage of some new technology, we may discover that it induces behavior which appears to cancel out the very reason for using it."[90]

One of the examples he uses compares the amount of injuries suffered in the game of football to those suffered in the game of rugby.[91] Those who have watched a rugby match may have cringed at the sight of some of the impacts between the players of opposing teams, while those same observers watching a football game think nothing of the dangerous effects that result from two 250 pound men running into each other at sprinter's speed. The difference, of course, is that the football players wear protective clothing while the rugby players do not. Tenner points out that the incidents of long-lasting violent injury are much higher in football than in rugby. He argues that the introduction of padding took away the concern for self preservation that rugby players use when deciding how physical to get with an opponent. The unintended consequence of padding, therefore, was that football players, dressed in their armor, do not temper their attacks because of the security they feel due to the protection of their equipment. Consequently, the increased energy in their contact outweighs the protection afforded by their padding, resulting in more injuries.[92]

UCAVs, too, will likely have their own unintended consequences. Though Tenner's work does not provide a simple formula or model for use in forecasting future consequences, it is incumbent upon the strategist to fully explore the realm of the

---

[90] Edward Tenner, *Why Things Bite Back: Technology and the Revenge of Unintended Consequences,* (New York, NY: Alfred A. Knopf, 1996), 1-5.

[91] Other examples concern the introduction of kudzu vine and technology driven devices. Kudzu vine was introduced in the 1950s as a way to prevent erosion. It is now a huge problem in the south because it is so vigorous and grows anywhere that it overtakes and kills off more valuable trees in its rapid spread. Remote control technology (such as self-propelled lawnmowers, etc.) may have had the unintended consequence of producing a society troubled by obesity due to the sedentary nature that the technologies allow.

[92] Tenner, 298-299.

possible when dealing with matters of national security. Failing to do so can result in a technology that might be too politically, morally, or socialy risky to use.

Thus far, this paper has addressed political, moral, and social issues separately and in isolation. However, when discussing the range of possible unintended consequences of extensive UCAV employment it is difficult to continue to do so due to the complexity and interdependence of these three areas. In the discussion that follows, the reader will see a convergence of political, moral, and social elements.

Increased Chance of War

As discussed in chapter one, UCAV employment, though not totally eliminating combatant risk in warfare, may make war "more palatable and possibly ubiquitous."[93] If the United States sees that it can fight wars with little loss of human life, will it be more prone to react to all differences of opinion through the use of force? Diplomacy may begin to take a back seat as force becomes easier to use, resulting in more conflict rather than less. This is especially true given that UCAVs offer the politician a quick and easy tool to use for coercive or signaling purposes. One would expect to see more one time or short duration strategic strikes given that UCAVs reduce much of the political baggage associated with conflict.

Expectations also play into this. As warfare becomes increasingly cleaner for Americans, expectations of casualties will naturally continue to decrease as Americans continue to see war as a video game.[94] Accustomed to sterile wars, the public may be more inclined to support a president who chooses force first, thus removing some of the restraint from the politician. Additionally, wars fought in this manner may reduce or remove some of the restraint that the military has towards war. That restraint comes from knowing and seeing the sometimes awful effects of what they are tasked to do. Take that restraint away and you risk having soldiers that are more eager to resort to force first as well, or at the very least, are slow to caution their political leaders against going to war.

Indeed, it has been argued that this is precisely what is happening today. According to Andrew Bacevich, the American reluctance to enter into war started its

---

[93] D. Keith Shurtleff, "The Effects of Technology on Our Humanity," *Parameters* Summer 2002, 103.
[94] Michael Ignatieff, *Virtual War: Kosovo and Beyond*, (New York, NY: Metropolitan Books, 2000), 168.

decline during the 1990's as the image of war as a brutal and barbaric act began to change. World War II, Vietnam, and Korea typified Americans expectations of war. After Desert Storm, however, America's perception of the brutality of war began to change. War was now cleaner and therefore more acceptable to the American public. This change made it easier for the United States to resort to force first rather than last as war has "become almost an annual event."[95] UCAVs, Bacevich would undoubtedly agree, have only accelerated that trend.

Alternatively, some may argue that the ability to ruthlessly prosecute war with no loss of American life is a great advantage and may have no down side. The United States, in a quest to spread democracy throughout the world, may be morally obligated to continually refine its means of waging war so that it can "kill precisely before [its] enemies kill indiscriminately."[96] Moreover, this capability can serve to send a signal to those who would challenge the United States. Potential adversaries may decide that it is hopeless to fight a nation that can conduct war without risk to its combatants. Therefore, they will seek peace and acquiesce to virtually all U.S. demands, resulting in a better state of peace throughout the world. The ability to conduct war with UCAVs, therefore, may actually reduce the incidence of war.

The likelihood of either scenario – an increased or decreased chance of war – depends on how prolific UCAVs become. If the military uses UCAVs in limited roles, performing only the "dull, dangerous, and dirty" tasks as outlined in the UAV roadmap, then there is little need for concern. They will be just another arrow in the military's quiver. However, if UCAVs eventually become a one-for-one replacement of manned aircraft, the evolution of war to the status of a video game may indeed remove or greatly reduce society's reluctance to go to war, making it more likely and widespread. If this is indeed the case and wars become more prolific, extensive UCAV employment may ultimately challenge the Just War ideals of maximizing good and minimizing evil.

However, even in the case of the latter, the effects will likely be short-lived. The second scenario assumes that future adversaries will be unable to counter American

---

[95] Andrew J. Bacevich, *The New American Militarism: How Americans are Seduced by War,* (Oxford: Oxford University Press, 2005), 19-21.
[96] David Skinner, "The New Face of War" *The New Atlantis: A Journal of Technology and Society,* Summer 2003, 7.

technological advances. If this were the case, it would mean a fundamental change in the dialectic of the ways and methods of fighting wars that has existed for centuries. In the long run, future enemies will likely find a way around the UCAV's advantages, bringing back the restraints that keep war within its moral bounds.

Reduced Public Support during War

Extensive UCAV use may make it difficult for the politician to *maintain* public support for long-term or complex wars fought predominantly with unmanned air assets if victory is not quick or decisive. There are several reasons for this. First, with no airmen in harm's way, politicians would not be able to rally support for an unpopular war by appealing to the emotions of an unsupportive populace. Without airmen deployed to foreign countries to fight America's battles, there will be little need for gestures such as yellow ribbons and service flags signifying public support. No longer would politicians be able to ask the public or a dissatisfied Congress to "support the troops in harms way."[97]

Additionally, the idea of staying committed to the cause so that "the troops will not have died in vain" would no longer apply. While this may sound counterintuitive, casualty aversion studies show that up to in certain point public support remains steady and sometimes increases as casualties mount because of a perceived need to justify past deaths and casualties through achievement of the goals of the war.[98] In economics, this is referred to as sunk cost theory. Economics teaches that businesses and individuals should not factor in past sunk costs when making future decisions, such as a decision to sell poorly performing stock. Generally, an individual tends to hold on to shares of stock even as the price continues to plummet, in the hopes of a future price increase; emotions rather than logic shape individual behavior. This pattern of behavior is even more prevalent when the sunk costs are measured in terms of lives instead of dollars.[99] Operations consisting predominately of UCAVs pull these options off the table.

---

[97] Woodley, 103.

[98] Edward N. Luttwak, *Strategy: The Logic of War and Peace,* (Cambridge, MA: The Belknap Press of Harvard University Press, 2001), 58.

[99] For more detailed analysis on the logic of decision making see: Amos Tversky and Daniel Kahneman, "The Framing of Decisions and the Psychology of Choice," *Science,* 211, no. 4481 (January 30, 1981), 453-458.

Therefore, the politician may no longer be able to resort to these emotional appeals to buoy support in complicated and messy wars fought predominately with unmanned aerial assets.

Second, public support might dwindle as a result of the increased attention given to enemy casualties. Since Vietnam, America has grown increasingly sensitive to all forms of casualties, even those to enemy combatants.[100] This sensitivity will only increase as extensive UCAV use contributes to a reduction in public concern over friendly losses, subsequently increasing the chance that the public might base their level of support on enemy casualty levels. This has already started to happen in recent conflicts. For instance, during the last days of Desert Storm, CNN footage of the "Highway of Death" showed the carnage inflicted by airpower on retreating Iraqi soldiers. The pictures provoked senior American officials to ask the military to halt such operations over concern that more footage like this would decrease the level of public support for an already successful war.[101]

Non-combatant casualties are even more polarizing. Though precision weaponry has allowed the United States to reduce the instances of collateral damage inherent in warfare, it has not eliminated it entirely. Precision airpower killed many civilians riding on a passenger train winding its way through Kosovo. It also took the lives of 65 Kosovar Albanians riding on a farm tractor escaping Serbian oppression. Aircraft recording equipment captured both instances, the results of which were released to many media outlets in an attempt to prove to the public that these were honest mistakes and not planned events. The leaders of each member country expressed concern over whether or not they could maintain public support in their respective countries in the aftermath of such incidences. Both events nearly broke-up the NATO alliance, putting the operation in jeopardy.[102]

---

[100] Harvey M. Sapolsky and Jeremy Shapiro, "Casualties, Technology, and America's Future Wars," *Parameters*, Summer 96, 1-3.

[101] Lawrence Freedman and Efraim Karsh, *The Gulf Conflict: Diplomacy and the New World Order*, (Princeton, NJ: Princeton University Press, 1993), 402-406.

[102] Luttwak, *Strategy*, 72-78.

Propaganda

Removing the human from the cockpit may only enhance the success with which the enemy could wage the war for hearts and minds through the use of propaganda. It is not difficult to imagine the enemy using pictures that depict ruthless but cowardly UCAV operators killing brave heroic soldiers willing to take on a technologically superior bully to increase their base of support while simultaneously reducing America's. Other pictures might show the same cowardly operator killing innocent women and children one minute and then rushing off to have lunch with his or her spouse and children the next.

The Vietnam War provides many examples of how the enemy used propaganda to reduce support for the war, but one stands out in particular. During one attack, an airplane dropped napalm on a village suspected of being an enemy stronghold. There were, of course, innocent non-combatants also in this village who suffered from the attack as well. Photographs quickly circulated around the world. Subsequently, the North Vietnamese government wasted no time in capitalizing on its propaganda value. They claimed that the incident was an example of a technologically superior country using modern technology to harm innocent civilians. Propaganda like this did much to damage America's reputation and credibility. Consequently, the American public as well as the international community reduced their support for the war, calling for American withdrawal from the region.[103] While manipulation for the purpose of propaganda is not unique to UCAVs, extensive use of unmanned assets will make it easier for the enemy to convince the world that the United States is nothing more than "a heartless craven society driven by technology."[104]

Rising Expectations

The next unintended consequence deals with expectations. Increasing UCAV employment in the future will undoubtedly save airmen's lives. The public may become accustomed to not having to deal with reports of airmen killed or missing in action, much like they already are when discussing air operations in general. For instance, in

---

[103] Phillip M. Taylor, *Munitions of the Mind: A History of Propaganda from the Ancient World to the Present Day*, (Manchester UK: Manchester University Press, 1995), 269-270.
[104] William Arkin, defense analyst and reporter, interview with author, 11 November 2004.

Operation Allied Force, NATO pilots flew thousands of sorties with zero friendly combat deaths, a miraculous feat given that the intensity of the Serbian air defense network. Additionally, during the 12 years that the military patrolled the Iraqi no-fly zones during Operations Northern and Southern Watch, not a single manned aircraft was shot down after numerous Iraqi attempts. Furthermore, the Air Force conducted operations in Afghanistan and Iraq with casualties only numbering in the single digits. UCAV employment will only increase the expectations that the military can conduct all air operations without casualties. However, that may not be possible.

Not all future missions may be well suited for UCAVs regardless of advances in unmanned vehicle technology. For instance, UCAV operators argue that close air support (CAS) missions that require a high degree of terminal attack control will continue to require the presence of manned aircraft such as the A-10 or AC-130, due to the close proximity of troops and dynamic battlefield conditions.[105] Their argument is based on the idea that even the most advanced UCAVs will not be able to react to the fast-paced and high-intensity nature of CAS missions where friendly and enemy forces are in close contact because it is more difficult for the UCAV operator to understand the true urgency and sense of this type of mission when he or she is not actually on the scene.[106]

These missions carry with them a great deal of risk for the aircrew. The nature of the mission requires the pilot to operate at low altitude, close to the ground where they are exposed to easily-obtained man portable air defense systems. There is a good chance that some aircrew will be shot down and subsequently killed in action or taken prisoner. How will the public respond when they have been conditioned by the extensive use of UCAVs to believe that air war is risk free? Will they insist that military leaders never place airmen in harm's way simply because current technology allows us to take the man out of the cockpit? This might limit the way the military responds to or prosecutes future wars.

---

[105] The Joint community differentiates between the three types of CAS control – Type 1, 2 or 3 – based on the degree of coordination required between ground controllers and CAS aircraft. Friendly forces in close contact with the enemy and dynamic battlefield conditions require tighter control procedures in order to decrease the occurrences of fratricide. See *Joint Publication 3-09.3, Joint Tactics, Techniques and Procedures for Close Air Support*, 3 September 2003 for a more thorough discussion.
[106] Lt Col Gary Fabricus, interviewed by author, 14 January 2005.

Precision guided munitions (PGMs) offer insight into how expectations based on past success can hamper the military commander. Through the successful use of PGMs, the military has been able to reduce the amount of collateral damage that occurs during air campaigns. Unfortunately, this high standard has turned even small collateral damage occurrences, once considered an unfortunate consequence of war, into a major event. In the public's eyes no collateral damage is now the rule rather than the exception. Today, downed aircraft are also considered an unfortunate consequence of war, yet this will likely change as the public comes to expect zero casualties during air operations in the era of UCAVs.

Additionally, military legal experts like Charles Dunlap have argued that future international law may even require PGM use by those states that have that technology.[107] Consequently, this has forced military commanders to scrutinize each and every target in an effort to reduce the possibility that collateral damage will occur, often at the expense of mission effectiveness. Military commanders are often hesitant to approve attacks without the advice of military lawyers or consent from higher command authorities. This was seen early in the Afghanistan war. In one instance, commanders could have engaged top al Quaeda operatives with Hellfire equipped Predators 10 times, but a laborious approval process spurred by concerns of collateral damage prevented a single shot.[108] In much the same way, the public or even the politician may demand that the military rely solely on UCAVs for all future aerial combat action, even at the expense of military efficiency and effectiveness. The extensive employment of UCAVs may add to a growing list of unhelpful expectations.

Risk Transfer

The final unintended consequence deals with a concept called risk transfer. Remotely piloting a UCAV directly reduces risk for the operator. But does that risk just disappear? Or is it merely transferred somewhere else? Tenner addresses this idea, but calls it risk displacement. One example he uses to explain this notion involves a case

---

[107] Charles J. Dunlap Jr., *Technology and the 21ˢᵗ Century Battlfield: Recomplicating Moral Life for the Statesman and the Soldier*, (Carlisle PA: Strategic Studies Institute, U.S. Army War College, 1999), 17.
[108] Thomas E. Ricks, "Target Approval Delays Cost Air Force Key Hits," *Washington Post*, 18 November 2001.

study analyzing the susceptibility of children to polio based on their socioeconomic status. In an attempt to stave off infections, medical professionals suggested that parents scrub their children down often in order to maintain a high level of hygiene. At the time, middle and upper class parents were more inclined or more able to keep their children clean or "scrubbed down" than poor parents were. Yet due to immune system tolerances, this left the "cleaner" children more susceptible to polio than the "dirty" ones. Reducing the risk from one type of infection ended up increasing the risk from another. Tenner argues that risk from general infection did not disappear through good hygiene, but was merely displaced.[109]

The first law of thermodynamics, also known as the conservation of energy, provides an additional metaphor for understanding the idea of risk transfer. The law states that energy can not be created or destroyed, only modified in form. In much the same way, risk does not totally disappear; some of it is transferred to other entities. Admittedly, risk is not a physical property like a liquid, gas, or solid. Moreover, the transfer process is not a zero sum game. Still, it offers a useful metaphor for understanding the potential unintended consequences of using UCAVs in combat.

The first place risk may be transferred is directly back to the front line combatant and it is the enemy's asymmetric response that acts as the mechanism to transfer the risk. When two combatants stand across from each other on the battlefield, armed only with sabers, the risk that each combatant experiences is essentially equal. But if one of those combatants asymmetrically responds by exchanging his sword for a gun, he reduces his own risk while increasing the risk for the opponent. Risk is therefore transferred from the combatant armed with the gun to the one armed with the sword. The dialectic continues if the soldier that has seen his risk increased now responds by arming himself with a machine gun in an effort to either equalize or transfer the risk back to the opponent.

Such a process of transfer may also occur as UCAV operations become more prolific. UCAV operators are like the combatant armed with the gun; though he greatly reduces his risk by transferring it to the opponent, it may only be for a short time. The enemy will look for ways to transfer this risk directly back to the operator. One could

---

[109] Edward Tenner, *Why Things Bite Back: Technology and the Revenge of Unintended Consequences,* (New York, NY: Alfred A. Knopf, 1996), 270.

easily imagine a scenario where the enemy directly attacks UCAV operators in their command facilities in the United States or abroad. Operators are especially vulnerable to this kind of attack because of their expectations concerning the nature of their operating environment. They expect it to be safe and innocuous, and therefore are not prepared for this type of asymmetric attack.

Combat support forces are the second place to which risk can get transferred. If the enemy finds that they cannot transfer the risk directly back to the front-line combatants due to the protective steps that these forces take, they will seek the path of least resistance and focus their efforts on new, less protected and more vulnerable targets. In the case of UCAVs, the enemy might transfer the risk to UCAV support personnel. Current and future UCAVs will depend on a robust command and control infrastructure. Satellite uplink, downlink, and relay stations will be one part of this network. Military personnel or contractors will operate these systems inside and outside of the continental United States. They will do so in what many think will be a relatively safe or risk-free environment. However, that safety may not last long as the enemy will look for ways to counter UCAVs, directly targeting those who operate this infrastructure. OIF provides as a useful example.

In past conflict, the rear area has been considered a relatively safe place to work. But in Iraq, it became arguably as dangerous as the front line. The enemy, unable to effectively challenge the overwhelming combat power of the coalition as they made their run to Baghdad, focused their efforts on rear-echelon forces. The ambush and subsequent capture of Corporal Jessica Lynch's transportation unit was the first of several such incidents. Though these rear area forces are lawful combatants, expectations were that they operated in a relatively risk free environment and did not need the same level of protection or training as their front line counterparts. The military did not account for asymmetric enemy tactics and the resulting transfer of risk.

Both of these examples involve risk transfer to legal combatants. Of greater concern in the moral realm, however, is when risk is transferred to the non-combatant. Enemy responses to UCAV employment may transfer risk to unsuspecting non-combatants. Innocent civilians would likely suffer casualties as a result of collateral damage from enemy attacks on UCAV operators or infrastructure. Moreover, the enemy may choose the indirect approach by threatening or attacking civilians or even family members of those who operate the systems. Therefore, as the United States begins to fight some of its wars from within the boundaries of its borders, it may actually increase the risk to non-combatants. We may, in effect, be bringing the battle to the non-combatant in an attempt to reduce the risk to combatants.

This is happening in Iraq as the insurgency is attempting to undermine U.S. attempts to install a democratic Iraqi government. As the U.S. military responded to insurgent tactics by increasing the armor on its soldiers and their vehicles, the enemy again looked to transfer the risk elsewhere. Consequently, the insurgents changed their methods and targets and began terrorizing the local population. Once again, risk was transferred to a particularly vulnerable group that did not expect to be targeted. Any Iraqi that directly or indirectly supported U.S. efforts became a target. Suicide bombers attacked those standing in line to work for the new government either through involvement in security affairs or those involved with the democratic election process. Insurgents also surrounded themselves with non-combatants as they sought refuge in Mosques and houses of Iraqi civilians. They also kidnapped and beheaded civilian contractors. In this case, non-combatants were the next stop in the risk transfer process.

Kosovo provides an additional example of how this can happen at the macro level. In Kosovo, NATO chose to use a less risky air campaign to coerce Milosevic into stopping the ethnic cleansing in the region. The coalition ruled out ground forces over concerns that their use would result in many casualties. NATO worried that as casualties increased, coalition stability would suffer, eventually leading to a failed campaign. However, the air campaign initially did little to stop the atrocities. In fact, Milosevic increased the intensity of the cleansing during the early stages of the air campaign, unconcerned about an impending ground attack.[110] Forces that would have had to defend themselves against a NATO ground campaign were free to terrorize Kosovar non-combatants. Risk to NATO forces was, in essence, transferred to civilian non-combatants.

---

[110] Stephen T. Hosmer, *The Conflict Over Kosovo,* (Santa Monica, CA: RAND Corporation, 2001), xiv.

Thus, categorizing UCAV employment as a risk free method of warfare may be a misnomer. All methods of warfare or forms of maneuver have their costs.[111] Germane to the idea of risk transfer is that the reduction of risk to the operator may incur some political, moral, or social costs: political, in that the politician will have to explain to the public how increasing the distance between shooter and target is beneficial when doing so might result in unexpected casualties to both combatants and non-combatants; moral, in that actions taken by a nation that result in non-combatant casualties are deemed immoral if the nation has the ability to fight by other means, in this case, manned aircraft; and social, in that society may in the long run be adversely affected if they have reason to fear for their life during distant conflict even though they are seemingly protected by two vast oceans. At issue is whether these costs are worth the benefits derived.

Conclusion

This is not an argument against UCAV employment on political, moral, or social grounds. It is, however, an argument against blindly accepting technological developments without thinking about the deeper political, moral, and social consequences of such an action. Strategists must weigh the possible costs with the benefits gained from UCAV employment and constantly look for ways to reduce or mitigate these costs so that they do not become prohibitive.

Several possibilities exist towards this end. The first step requires that the military and the political establishment acknowledge, at least privately, the possible long-term effects of remote control war on decision making and policy to prevent war from becoming more prolific. Second, when prosecuting war with extensive use of unmanned air assets, politicians need to understand that it may become more difficult to appeal to the emotions of the public if the war becomes messy and drawn out. Politicians will still need to mount robust public relations campaigns even in the absence of friendly casualties. Third, the military must be ready to counter enemy propaganda aimed at U.S. technological superiority, showing that this new technology saves lives on all sides of the conflict. Fourth, the military should not oversell the abilities of UCAVs lest they create unrealistic expectations that may, in the end, limit military options and hamper the

---

[111] Edward N. Luttwak, *Strategy*, 5.

military effectiveness of future air campaigns. Fifth, the military should make every attempt to isolate UCAV operations to areas where few non-combatants work or live and ensure suitable force protection measures are established in these areas. This would reduce the chances of innocent civilians suffering the consequences of collateral damage from asymmetric attacks. Additionally, if the military continues on its present path, using civilian contractors in support of UCAV operations, they must ensure that the contractors know that they run the risk of being classified by the enemy as combatants, regardless of their civilian status. Finally, the military should closely guard the identities of UCAV operators to reduce the chance that their families become the target of enemy asymmetric attacks.

Conclusion

UCAVs are a very appealing option for politicians faced with use-of-force decisions and may indeed make them more inclined to resort to force first rather than last. However, the extent to which this occurs will likely depend on the type and nature of the conflict. When faced with short term strikes and raids and politically sensitive conflicts, both domestically and internationally, the UCAV's reduced forward basing requirements and reduced friendly casualty and POW chances may entice the politician to forgo diplomacy and choose force, or at least reduce reliance on the former. When these advantages are coupled with the UCAV's ability to persistently track, target, engage, and assess targets, the politician may feel that he is able to precisely control the employment of force, ratcheting coercive pressure up or down as needed. Yet this feeling of control is dangerously seductive. The danger is that he or she may tend to lose sight of larger objectives, or worse yet, forget that they have other instruments of power at their disposal.

Analysis of the moral implications of UCAVs shows that UCAVs do not violate the *jus in bello* principles of proportionality or discrimination. Additionally, they do not violate any preexisting rules concerning legal and moral weapons outlined in the LOAC. Consequently, more often than not, it is the way in which a weapon is used – and not the actual weapon itself – that determines its morality. In fact, UCAVs actually increase the military commander's ability to adhere to these principles in that they allow him to reduce risk to his forces, thereby fulfilling his moral obligations to own his troops. This ultimately contributes to Just War theory's desire to maximize good and minimize evil.

Chivalry and fairness are poor standards with which to judge the morality of new technology. Chivalry, as it was originally conceived, was only concerned with making warfare more palatable to the aristocratic elite and not for the average combatant. Warfare might have been even less bloody and indiscriminate for the average citizen and warrior had chivalry never existed in the first place. Furthermore, actions taken by a military commander are always fair if they are in accordance with the LOAC and Just War theory principles. Both concepts are misunderstood by both the military and the average citizen at large.

Simply removing operators from the battlefield does not reduce the need of an ethos to guide their behavior in warfare. The military ethos exists to provide guidelines and standards of conduct for the members of the military who are asked to kill the enemy regardless of distance from the enemy when the killing takes place. It is the end result of the military member's duty, the act of killing, which provides the impetus for the ethos. UCAVs do not remove this duty from the operator. He or she is still responsible for taking life on the battlefield and therefore subject to the military ethos. This interpretation of the ethos, however, is not necessarily the one that the Air Force as an institution has emphasized. The institution and its members are still focused on and tend to stress the *closing* characteristic of the ethos instead of the *killing* aspect. Therefore, Air Force leaders should stop focusing solely on the *closing* characteristic of the ethos. Changing the emphasis is especially important as technology continues to further remove the warrior from the battlefield. Instead, service leaders should acknowledge that some airmen will close with the enemy while others will not. Regardless of the nature of the weapon system, the end result is still the same. The ethos exists to provide guidelines to the airmen in carrying out his or her duties, duties that are unique and distinctive from the rest of society.

While UCAVs have many advantages, they do not come without costs in terms of possible unintended consequences, consequences that cross over political, moral, and social stratum. First, as America becomes increasingly used to clean or sterile war with regards to friendly casualties, UCAVs may remove, or at least reduce, some of the restraining forces that have kept force within its bounds thereby making war "more palatable and possibly ubiquitous."[112]

Second, politicians may find it harder to maintain public support for a war that lasts longer than expected even if experiencing few friendly casualties. The politician will no longer be able to appeal to the emotions of the public to support the troops in harm's way or stay committed to the cause so that the troops will not have died in vain.[113]

[112] D. Keith Shurtleff, "The Effects of Technology on Our Humanity," *Parameters* Summer 2002, 103.
[113] R. Ross Woodley, "Unmanned Aerial Warfare, Strategic Help or Hindrance," (master's thesis, School of Advanced Air and Space Studies, Air University, Maxwell Air Force Base, AL, 2000), 103.

Moreover, lack of friendly casualties will likely result in increased emphasis on enemy casualties, both non-combatant as well as combatant.

Third, removing the human from the cockpit may also enhance the success with which the enemy can wage a war for hearts and minds through the use of propaganda aimed at showing how the United States is unable or unwilling to challenge the enemy directly on the battlefield. America's reliance on remote-control warfare may sway international opinion against the United States as it conducts war without risk to its own airmen.

Fourth, the notion that politicians may end up losing some of the flexibility that UCAVs were designed to offer as increasing expectations may result in tying the hands of the decision maker. Increasing UCAV employment in the future will undoubtedly save airmen's lives and the public will become accustom to not having to deal with reports of airmen killed or missing in action. Yet not all future missions will be suitable to UCAVs. Consequently, when the first airmen is shot down and killed or held POW, the public may question why UCAVs were not used, subsequently demanding their use in all aerial combat operations.

Finally, and of greater concern to the strategist, is the idea that UCAVs might actually increase the overall danger to unsuspecting combatants and non-combatants in a risk transfer process. We should not assume that reducing risk through distant warfare removes risk entirely from the battlefield. Metaphorically speaking, risk, like energy or momentum, has to go somewhere. If it ends up back on the legal combatant, there is little question as to the morality of the UCAV. However, risk transferred to the non-combatant is great cause for concern and does fall under the realm of moral implications.

The key take-away from this analysis is that the strategist should delve deeper into the political, moral, and social implications of new technologies. Stopping the analytical process after determining that a technology passes simple tests of morality and legality, for instance, is not going far enough. This work has pushed beyond this starting point by framing the questions that must be asked and providing some preliminary answers, but it has merely scratched the surface of the possible political, moral, and social implications of UCAV employment. As UCAVs proliferate, some of these implications may prove true, while others may not. Only time will tell what the actual short and long term implications of UCAV employment will be. Undeniably, there will be implications arising from their use, implications that will affect both development and strategy. Regardless of whether the effects are beneficial or not, strategists owe it to themselves and the leaders they serve to analyze the effects that the weapons of war have on the formulation of strategy.

This paper, therefore, serves as a starting point, not a starting point in the sense that these are the first implications in a long list, but a starting point in thinking about the political, moral, and social implications of all future military technologies. Strategists must understand that they cannot merely examine new technologies against a single issue in isolation. The implications of new technology cut across political, moral, and social bounds. Each issue is highly complex and interdependent.

In the end, strategists would be wise to remember Clausewitz's thoughts on weapons technology. He correctly noted that weapons do not change the eternal logic of war, only the evolving grammar. This piece of wisdom is as true in the age of the UCAV as it was in the age of the musket.

BIBLIOGRAPHY

Bacevich, Andrew. "Morality and High Technology." *National Interest*, Fall 1996.

Bender, Brian. "Attacking Iraq from a Nevada Computer.ty" *Boston Globe*, section A6, 3 April 2005.

Berkowitz, Bruce. *The New Face of War.* New York, NY: The Free Press, 2003.

Boot, Max. "The New American Way of War." *Foreign Affairs*, July/August 2003.

Buckingham, David W. "The Warrior Ethos." Master's thesis, Naval War College, 1999.

Builder, Carl H. *The Masks of War: American Military Styles in Strategy and Analysis.* Baltimore, MD: The RAND Corporation, The Johns Hopkins University Press, 1989.

—————— *The Icarus Syndrome: The Role of Airpower Theory in the Evolution and Fate of the U.S. Air Force.* New Brunswick, NJ: The RAND Corporation, Transaction Publishers, 1994.

Bush, George W. "Presidential Press Conference." 28 October 2003.

Butler, Amy. "DOD, State Dept Debate Whether 'Weaponized' UAV Would Violate Treaty." *Inside the Air Force*, 8 Dec 2000.

Cameron, Bud. "When Robots Kill." Thesis, Canadian Forces College, 2003.

CBS News. "US Charges Cole Role Players." *CBS News.com.* On-line. CBS News.com, 9 February 2005.

Christopher, Paul. *The Ethics of War and Peace: An Introduction to Legal and Moral Issues.* 3rd ed. Upper Saddle River, NJ, Prentice Hall, 2004.

Coates, A.J., *The Ethics of War.* New York: Manchester University Press, 1997.

Cohen, Eliot A., "The Mystique of U.S. Air Power." *Foreign Affairs,* January/February 1994.

—————— "A Revolution in Warfare." *Foreign Affairs.* March/April 1996.

Cohen, Eric. "The New Politics of Technology." *The New Atlantis: A Journal of Technology and Society.* Spring 2003.

Coker, Christopher. *Humane Warfare.* London, Routledge, 2001.

—————— *Waging War Without Warriors: The Changing Culture of Military Conflict.* Boulder, CO: Lynne Rienner Publishers, 2002..

Colucci, Frank. "Air Force Refines Training Programs for UAV Operators." National Defense 88, no. 606 (May2004): 37-39.

Congressional Research Service *Operation Enduring Freedom: Foreign Pledges of Military Intelligence Support.* Report for Congress. Washington, D.C.: October 17, 2001.

Congressional Research Service, *Unmanned Aerial Vehicles: Background and Issues for Congress.* Report for Congress. Washington, D.C.: April 25, 2003.

Coppieters, Bruno. *Moral Constraints on War: Principles and Cases*, Edited by Bruno Coppieters and Nick Fotion. Lanham, MD, Lexington Books, 2002.

Crawley, Vince and Amy Svitak. "UAV Strike Raises Moral Questions." *Air Force Times* 63 no. 17 (18 Nov 2002): 16.

Donovan, Aine. *Ethics for Military Leaders.* 2nd ed. Needham Heights, MA, Pearson Custom Publishing, 1999.

Douglas, Mark. "Changing the Rules: Just War Theory in the Twenty-First Century." *Theology Today,* 59: 529-546, January 2003.

Dunlap, Charles J. Jr. *Technology and the 21st Century Battlefield: Recomplicating Moral Life for the Statesman and the Soldier.* Strategic Studies Institute, 15 January 1999.

Erhard, Thomas P. "Unmanned Aerial Vehicles in the United States Armed Services: A Comparative Study of Weapon System Innovation." PhD diss., Johns Hopkins University, June 2000.

French, Shannon E. *The Code of the Warrior: Exploring Warrior Values Past and Present.* Lanham, MD: Rowman & Littlefield Publishers, Inc., 2003.

Gentry. "Military Force in an Age of Cowardice." *Washington Quarterly.* Autumn 1998.

Gingras, Jeffrey L. and Ruby, Tomislav Z. Morality and Modern Air War. *JFQ: Joint Force Quarterly* 25:107-111 Summer 2000.

"Global Hawk." *Air Force Link.* On-line. www.af.mil/factsheets/factsheet.asp?fsID=175, 27 Feb 2005.

Goldman, Michael W., ed. *The Military Commander and the Law.* Maxwell Air Force Base, AL: Air Force Judge Advocate General Press. 2004.

Grant, Rebecca. "Khobar Towers." *Air Force Magazine,* June 1998. *Air Force Magazine Online.* www.afa.org.magazine/jine1998/0698khobar.asp, 27 March 2005.

Gray, Colin S. *Weapons for Strategic Effect,* 2001, On-line. www.au.af.mil/au/awc/awcgate/cst/csat21.pdf

Grey, Stephen. "Pilotless Strikes on Iraq by RAF." *London Sunday Times,* 3 October 2004.

Grossman, Dave. *On Killing: The Psychological Cost of Learning to Kill in War and Society.* New York: Little, Brown and Company, 1996.

Hackett, Sir John. *The Profession of Arms.* New York, NY: Macmillan Publishing Company, 1983.

Hanson, Victor Davis. "Military Technology and American Culture" *The New Atlantis: A Journal of Technology and Society,* Spring 2003.

Huntington, Samuel P. *The Soldier and the State: The Theory and Politics of Civil-Military Relations.* Cambridge, MA: The Belknap Press of Harvard University Press, 1957.

Ignatieff, Michael. *Virtual War: Kosovo and Beyond.* New York, NY: Metropolitan Books, 2000.

Johnson, James T. *Morality and Contemporary Warfare.* Yale: Vail-Ballou Press, 1999.

Johnson, Michael W. *Just-War Theory and Future Warfare.* Master's thesis, U.S. Army Command and General Staff College, 1999.

Joint Chiefs of Staff. *Joint Vision 2020.* Washington, D.C.: Office of the Chairman of the Joint Chiefs of Staff (Director for Strategic Plans and Policy, J5, Strategy Division), June 2000.

Lang, Anthony F.; Albert C. Pierce; Joel H. Rosenthal. *Ethics and the Future of Conflict: Lesson from the 1990s.* Upper Saddle River, NJ: Pearson/Prentice Hall, 2004.

Larson, Eric V. *Casualties and Consensus: The Historical Role of Casualties in Domestic Support for U.S. Military Operations.* Santa Monica, CA.: RAND Corporation, 1996.

Lazarski, Anthony J. "Legal Implications of UCAVs," *Air and Space Power Chronicles,* 27 Mar 2001.

Lockhart, Joe. "White House Press Briefing on Operation Allied Force." 21 May 1999. On-line. www.clinton foundation.org/legacy/052199.

Luttwak, Edward N. "A Post-Heroic Military Policy." *Foreign Affairs,* May/June 1995, 42.

————————— *Strategy,* Cambridge, MA: The Belknap Press, 1987.

Lynn, John A. *Battle: A History of Combat and Culture.* Cambridge, MA: Westview Press, 2003.

Mapel, David R., 1996. "Realism and the Ethics of War and Peace." In *The Ethics of War and Peace: Secular and Religious Perspectives.* Edited by Terry Nardin. Princeton, NJ: Princeton University Press. 54-77.

Matthews, Mark T. "A Search for Warriors: The Effects of Technology on the Air Force Ethos." Master's thesis, Air War College, Maxwell Air Force Base, AL, 1997.

McMichael, William H. "Punishment Meted Out for Six in Grounding of Submarine," *Navy Times,* 22 Mar 2005. *Navy Times.com.* On-line. 26 March 2005.

Meileander, Gilbert. "War and Techne," *The New Atlantis: A Journal of Technology and Society,* Summer 2003.

Meilinger, Phillip S. "A Matter of Precision: Why Air Power may be more Humane than Sanctions." *Foreign Policy,* Mar/Apr 2001, 78-79.

Moore, David W. "Eight of 10 Americans Support Ground War in Afghanistan." *The Gallup Organization,* 1 November 2001. *Gallup Online.* On-line. www.gallup.com/content/login.aspx?ci=5029. 5 February 2005.

Mueller, Karl. "Politics, Death, and Morality in US Foreign Policy." *Aerospace Journal,* Summer 2000.

Pae, Peter. "Air Force Wants Big Boost in Predator Fleet." *Los Angeles Times,* 19 Mar 2005. *Current News Early Bird.* On-line at http://ebird.afis.osd.mil/ 20 Mar 2005.

Peters, John. "A Potential Vulnerability of Precision-Strike Warfare." *Orbis,* Summer 2004.

Peters, Ralph. "A Revolution in Military Ethics." *Parameter,.* Summer 1996.

Ramsey, Paul. The *Just War: Force and Political Responsibility.* Lanham, MD: Rowman & Littlefield, 2002.

Ricks, Thomas E. "Target Approval Delay Costs Air Force Key Hits." *Washington Post,* 18 November 2001.

Robinson, Bruce. *The Pals Battalions in World War One.* On-line. www.bbc.co.uk/history/war/wwone/pals_01.shtml. 5 February 2005.

Roche, James G. "The Centennial Airmen – A New Generation of Air and Space Leaders." U.S. Air Force Academy Commencement Address, 28 May 2003.

Ruben, Charles T. "Artificial Intelligence and Human Nature." *The New Atlantis: A Journal of Technology and Society,* Spring 2000.

Sapolsky, Harvey M. and Jeremy Shapiro. "Casualties, Technology, and America's Future Wars." *Parameter,.* Summer 1996.

Simms, John R. *Shackled by Perceptions: America's Desire for Bloodless Intervention,* Master's thesis, School for Advanced Air and Space Studies, Air University, Maxwell Air Force Base, AL, 1997.

Shlapak, David A., et al. "*A Global Access Strategy for the U.S. Air Force.* Project Air Force. Santa Monica, CA: Rand Corporation, 2002.

Shurtleff, D. Keith. "The Effects of Technology on Our Humanity." *Parameters,* Summer 2002.

Skinner, David. "The New Face of War" *The New Atlantis: A Journal of Technology and Society,* Summer 2003.

Soller, Daniel, E. *Operational Ethics: Just War and Implications for Contemporary American Warfare.* Master's thesis, School of Advanced Military Studies, Army Command in General Staff College, Fort Leavenworth, KS, 2003.

Starr, Barbara. "U.S. to Move Operations from Saudi Base." *CNN.com,* 29 April 2003. *CNN News* On-line. www.cnn.com/2003/worldmeast/04/29/sprj.irq.saudi.us/. 5 February 2005.

Squeo, Ann Marie. "Top Guns Grounded: Pilots Fume at Duty on Unmanned Aircraft— Fliers Used to F-16s Now sit in Windowless Cubicles; Shot Down at the Arcade." *Wall Street Journal,* 29 April 2003.

Taylor, Phillip M. *Munitions of the Mind: A History of Propaganda from the Ancient World to the Present Day.* Manchester, UK: Manchester University Press, 1995.

Temes, Peter S. *The Just War: An American Reflection on the Morality of War in Our Time.* Chicago: Ivan R. Dee, 2003.

Tenner, Edward. *Why Things Bite Back: Technology and Unintended Consequences.* New York, NY: Alfred A. Knopf, 1996.

Thomas, Evan, and Mark Hosenball. "The Opening Shot." *Newsweek,* 18 November 2002, 48.

Tirpak, John A. "Airpower and 'The Long War', Four-Star Forum: Eyes on the Future," *Air Force Magazine,* 87 (November 2004): 75-83.

Tversky, Amos and Daniel Kahneman. "The Framing of Decisions and the Psychology of Choice." *Science,* 211, no. 4481 (January 30, 1981):453-458.

United States Army. *Army Transformation Road Map.* Washington D.C., 2003.

U.S. Air Force. *The U.S. Air Force Transformation Flight Plan.* Washington, D.C.: Headquarters United States Air Force (XPXC Future Concepts and Transformation Division), November 2003.

U.S. Department of Defense. *Joint Publication 1-02: Dictionary of Military and Associated Terms.* Washington, D.C.: Office of Secretary of Defense, 30 Nov 2004.

U.S. Department of Defense. *Unmanned Aerial Vehicles Roadmap: 2002-2027.* Washington, D.C.: Office of the Secretary of Defense (Acquisition, Technology, & Logistics), December 2002.

United States Navy. *Naval Transformation Road Map 2003.* Washington D.C..

US Senate. *National Defense Authorization Act for Fiscal Year 2001.* 106th Cong., 2nd sess., S.R. 2549. *Congressional Record,* no. 543, (12 May 2000).

Walzer, Michael. *Just and Unjust Wars.* New York: Basic Books, 1977.

_____. "Kosovo" *Dissent* 46, Summer 1999, pp 5-7.

Wetterhahn, Ralph. *The Last Battle: The Mayaguez Incident and the End of the Vietnam War.* New York: Carroll & Graf Publishers, Inc., 2001.

Woodley, R. Ross. "Unmanned Aerial Warfare, Strategic Help or Hindrance." Master's thesis, School of Advanced Air and Space Studies, Air University, Maxwell Air Force Base, AL, 2000.